SKETCHES OF NATURE

SKETCHES OF NATURE

A GENETICIST'S LOOK AT THE BIOLOGICAL WORLD DURING A GOLDEN ERA FOR MOLECULAR ECOLOGY

JOHN C. AVISE

Department of Ecology and Evolutionary Biology,
University of California at Irvine, CA, USA

AMSTERDAM • BOSTON • HEIDELBERG • LONDON
NEW YORK • OXFORD • PARIS • SAN DIEGO
SAN FRANCISCO • SINGAPORE • SYDNEY • TOKYO

Academic Press is an imprint of Elsevier

Academic Press is an imprint of Elsevier
125, London Wall, EC2Y 5AS
525 B Street, Suite 1800, San Diego, CA 92101-4495, USA
225 Wyman Street, Waltham, MA 02451, USA
The Boulevard, Langford Lane, Kidlington, Oxford OX5 1GB, UK

Notices
Knowledge and best practice in this field are constantly changing. As new research and experience broaden our
understanding, changes in research methods, professional practices, or medical treatment may become necessary.

Practitioners and researchers may always rely on their own experience and knowledge in evaluating and using
any information, methods, compounds, or experiments described herein. In using such information or methods
they should be mindful of their own safety and the safety of others, including parties for whom they have a
professional responsibility.

To the fullest extent of the law, neither the Publisher nor the authors, contributors, or editors, assume any liability
for any injury and/or damage to persons or property as a matter of products liability, negligence or otherwise,
or from any use or operation of any methods, products, instructions, or ideas contained in the material herein.

ISBN 978-0-12-801945-0

Library of Congress Cataloging-in-Publication Data
A catalog record for this book is available from the Library of Congress

British Library Cataloguing-in-Publication Data
A catalogue record for this book is available from the British Library

For information on all Academic Press publications
visit our website at http://store.elsevier.com

 Working together
to grow libraries in
ELSEVIER | **Book Aid** International | developing countries

www.elsevier.com • www.bookaid.org

Publisher: Sara Tenney
Acquisition Editor: Kristi A. S. Gomez
Editorial Project Manager: Pat Gonzalez
Production Project Manager: Melissa Read
Designer: Matthew Limbert

Printed and bound in the United States of America

Contents

Preface

The final third of the twentieth century was a Golden Age for evolutionary ecology. It was a time when the newfound tools of molecular biology were first being deployed to explore the remarkable genetic diversity of the organic world. In the initial 100 years following Darwin (1859) and Mendel (1865), natural historians had been obliged to confine their attention to organismal phenotypes (such as morphologies and behaviors) whose genetic bases could be surmised only loosely. The molecular revolution opened entirely new windows on nature. Beginning in the mid-1960s, a succession of protein-level assays and DNA techniques revealed plethoras of genetic polymorphisms that biologists soon exploited to address nature's secrets. Molecular markers do not supplant traditional field observations; when used to best effect they supplement the latter in a powerful synergy that has become the hallmark of modern-day evolutionary genetics.

This book is an attempt to capture the spirit of the Golden Age in molecular ecology and evolution, by way of example. It presents published summaries of more than 200 scientific papers from the laboratory of John C. Avise (JCA), spread over a span of more than four decades following the dawn of the molecular period in natural history. The "sketches" are of two types: (i) the written abstracts themselves, which provide glimpses into one laboratory's experience with the development and application of molecular markers to explore a wide variety of species in the wild and (ii) lovely line drawings by Trudy Nicholson, one of the world's outstanding natural history artists. We hope that this book captures the essence of this special era in evolutionary biology in an artistic manner that is both educational and entertaining.

Introduction

Nucleic acids and the proteins they specify are universal to life on Earth, but each species, population, and individual carries its own idiosyncratic genetic versions. In the latter third of the 20th century, evolutionary geneticists began using the laboratory tools of molecular biology to decipher and compare these genomic texts from a wide variety of organisms. A net result was the explosive growth of molecular ecology and evolution, a field that can have either of two research foci: the molecules *per se*, or the organisms that house them. The author's (JCA's) career has been devoted to the latter arena, viewing the behavior and evolution of organisms in the wild through the novel lenses provided by polymorphic proteins and DNA markers.

Never again will there be a scientific era quite like during the meteoric rise of molecular ecology in the decades following the 1950s. It was the time of a stunning transition from traditional organismal biology to modern genomic approaches. Most evolutionary biologists at that time had little background or knowledge of molecular genetics, having been trained instead in organismal fields such as systematics, natural history, ecology, or one of the many creature "ologies" such as ichthyology, ornithology, mammalogy, herpetology, or invertebrate biology. Thus, when the first of many molecular techniques—protein electrophoresis—was introduced to population biology in the mid-1960s, it triggered a scientific revolution that was both exhilarating and somewhat threatening to many professional biologists. Would the new molecular data overturn some of the long-accumulated wisdom of the "modern evolutionary synthesis, " or would it bolster the latter and eventually become incorporated into it? This was a time of expanding scientific horizons and frontier explorations into a brave new molecular realm. It was a time when the entire world of nature was first opened to close genetic scrutiny. This book is my attempt to capture the intellectual essence of this exciting period in the history of ecological and evolutionary genetics, as seen through the lens of one laboratory's experience in applying various emerging molecular technologies to a diverse array of organisms from natural populations.

The Golden Age for ecological and evolutionary genetics, which spanned from the 1960s through the early 2000s, itself consisted of several recognizable sub-eras each motivated by introduction of a novel molecular approach to natural population analysis. First came the allozyme era of the 1960s and 1970s, which capitalized upon the plethora of Mendelian polymorphisms that were newly revealed using multi-locus starch-gel protein electrophoresis (SGE). These nuclear genetic polymorphisms found application in a wide variety of topics ranging from genetic parentage assessments, to population structure analyses, to phylogenetic appraisals that proved relevant for taxonomy and systematics. Next came the mitochondrial (mt) DNA heyday of the 1970s and 1980s, which followed the discovery of restriction endonucleases and their application to reveal restriction fragment length polymorphisms (RFLPs) in the cytoplasmically housed and maternally inherited mtDNA genome. Applications included phylogeographic appraisals of numerous animal species, as well as refined genetic dissections of hybridization and introgression phenomena. Then came the microsatellite era beginning in the late 1980s and 1990s. With respect to magnitude of genetic variation, microsatellite polymorphisms are like allozyme polymorphisms on steroids. Microsatellite loci often display dozens to scores of distinguishable alleles and extraordinarily high heterozygosities, so these nuclear Mendelian markers proved to be ideal for genetic assessments of microevolutionary phenomena such as clonal recognition, paternity and maternity assignments, and genetic appraisals of mating systems in nature. In the late 1990s through the 2000s, and continuing today, a succession of dramatic improvements in DNA sequencing ushered in the genomics era, in which the analysis and interpretation of ubiquitous molecular variation have become even more challenging than data acquisition itself. It remains to be seen how this genomics era will alter the trajectories of molecular ecology and molecular evolution. Throughout the Golden Age of the late twentieth century, the acquisition of extensive DNA sequences had always been the Holy-Grail quest for population and evolutionary geneticists. Now that such molecular data have become readily available, what will the next half-century bring? Will the new data merely add refinements to our understanding of nature, or are we perhaps at the dawn of what will someday be perceived as a second Golden Age in molecular evolutionary genetics?

I wrestled with several alternative ways to organize the abstracts presented in this book. One approach would have emphasized the succession of laboratory techniques employed (as described above). The first part of JCA's career centered on allozyme (protein-electrophoretic) and chromosomal (karyotypic) methods; the second part shifted to an emphasis on mitochondrial (mt) DNA; and a third segment mostly shifted to assays of highly polymorphic microsatellite loci. Another approach would have been to organize the chapters on the basis of topical areas (which are partially intertwined with the laboratory methods deployed). JCA and his students and colleagues have studied many biological subjects including genetic parentage in local populations, geographic population structure and gene flow, speciation, hybridization and introgression, phylogeography, and interspecific phylogenetics. Ideally, each such topic should be matched to its most suitable suite of molecular markers.

In the end I decided to arrange the abstracts according to the organismal groups, reasoning that this format would best serve most readers. Most natural historians are interested primarily in some particular group of creatures, such as birds, marine turtles, freshwater fishes, or invertebrate animals. The chapter headings reflect such animal amalgamations. Within each chapter, abstracts are arranged mostly chronologically, generally reflecting both the temporal succession of laboratory techniques—and the conceptual train of thought—underlying JCA's research program on each organismal assemblage. Such temporal ordering should be of special interest to biology students and science historians. Some of the earliest papers in this collection will seem simple or even quaint by modern genomic standards, but that is precisely one point of this historical exercise. Although this book encapsulates JCA's research in an abbreviated format, interested readers can find the electronic pdf for any full-length paper at the following website address: http://faculty.sites.uci.edu/johncavise/publications/.

Here is a list of this book's top ten epistemological rationales. This book will:

1. highlight the fact that protein and DNA markers are universal to life on Earth;
2. demonstrate the power of molecular technologies in ecology and evolution;
3. emphasize the inherently comparative nature of molecular information;
4. trace the history of molecular ecology and evolution via personal experiences;
5. illustrate a full panoply of research topics in molecular natural history;
6. advertise and reminisce a magical time in the history of biological exploration;
7. show how two disparate fields—molecular genetics and natural history—can be wed;
8. revel in the exuberant molecular genetic diversity of natural populations and species;
9. introduce a new format for encapsulating the scientific career of a prominent researcher; and
10. showcase animal drawings by one of the world's greatest natural history artists.

Overall, my hope is that this collection of abstracts, anecdotes, and addenda, accompanied by relevant organismal drawings, will offer readers the intellectual gist—unencumbered by the arcane details that characterize full-length scientific articles—of one laboratory's research experiences during a Golden Era for molecular ecology and evolution.

Collecting the Animals

At the outset of any research project in molecular ecology and evolution, a major hurdle is first getting the animals (or their relevant tissues) into the laboratory. In other words, before any protein or DNA analyses can be conducted "at the bench," the animals must first have been collected from nature. Although this point may seem self-evident, it is worth emphasizing because collecting can be the most challenging aspect of an entire research enterprise, perhaps necessitating arduous fieldwork (sometimes in remote or dangerous locations) conducted over many days, weeks, or even months. Furthermore, the field technique(s) employed in a given project can be quite inventive, because each method must be finely tailored to the idiosyncratic habits of the particular organisms being sought. JCA is never happier than when "in the field" collecting research animals, preparatory to the laboratory drudgery that inevitably will follow. The author and his students literally have roamed the world in search of their quarry for various molecular genetic projects. This chapter will merely introduce some of the many field techniques that JCA's personnel have employed at one time or another to collect specimens for this research. Many of these collecting methods may be surprising to laboratory biologists or indeed to any readers unaccustomed to fieldwork with natural animal populations.

Fishing. Standard methods for collecting fish include hook-and-line angling, seining, gill-netting, and electroshocking, all of which have been employed routinely by JCA, depending on the species and environmental setting. Marine catfish, sea basses, and largemouth bass are examples of large open water species that were often taken by standard hook-and-line. Seining is an effective and oft-used technique especially for abundant shallow water species (such as bluegill sunfish or various minnows) when the waterway is relatively unobstructed (such as along a sandy beach). The process requires two people, each of whom holds a pole between which is stretched a long (e.g., 50 foot × 4 foot) net that is dragged through the water, thereby capturing fish that are then hauled ashore. Gill-netting is similar except that the net is larger meshed and is strung out for a time between two stationary poles (or attached to floats). After several hours, the net is checked for any sought fish that may have become entangled in its mesh. (This procedure also helped us to catch diamondback terrapins for one genetic project.) Electroshocking is another useful approach for collecting many fishes (such as cutthroat trout or mottled sculpins) from fast-flowing rivers or other shallow bodies of freshwater that are heavily strewn with rocks or other debris. This method uses an electrical power generator (carried in a backback or on a boat) attached to two long electrodes that the operator dips into the water to shock any fish that are nearby. The stunned fish then float to the surface where they are scooped using hand-held dipnets. Of course the fishers must wear rubber gloves and waders, lest they too get inadvertently shocked during this noisy operation. On rare occasions, the author has also used rotenone to capture fish from small isolated or otherwise inaccessible bodies of water. When dripped into the water, this chemical suffocates the fish and causes them to float to the surface where they can be scooped up.

Several harvested fish species, such as menhaden, herring, and eels, were purchased "at the dock" from commercial fishers or were taken in collaboration with research personnel from museums or from federal or state departments of Fish and Wildlife. What follows are some additional or more specialized methods of fish collecting that have been employed for particular species in one or another research project in JCA's laboratory.

Cave fish. Obtaining troglobitic species for his Masters thesis required that JCA learn spelunking techniques. Over several weeks, the author practiced rappeling by rope and developed other caving skills that proved necessary for accessing pools in the pitch-black Mexican caves in which these eyeless little fishes reside.

Mangrove killifish. The inch-long members of this species often live in the bottom of crabs' burrows that are intertwined in a tangle of mangrove roots in mosquito-infested lagoons. We have extracted fish from these lairs using miniature baited hooks and a line attached to a small stick, and/or by using tiny dipnets.

Nest-tending species. In many fish species, males build and tend nests that may house dozens to hundreds of embryos upon which we have conducted genetic maternity and paternity analyses. Sometimes (as in sticklebacks) the nests are woven in vegetation above the substrate in shallow marine waters, and thus are readily visible and can be collected in a glass jar simply by snorkeling. Other nests (such as in sunfish) are plate-like depressions in the silty substrate of a lake or stream. These offer somewhat stiffer collecting challenges. In such cases, we first electroshocked the nest attendant from his nest, before sampling the embryos either by plunging a glass jar into specific sites in the nest, or by gently scooping the entire next and its contents into a large plastic bin. The embryos are then laboriously sorted from the detritus and plucked by eyedropper under a magnifying glass.

Grunion. These marine fish come ashore only briefly, on high-tide nights a few times a year, to lay their eggs on wave-washed sandy beaches. We collected spawning specimens, plus fertilized eggs from their nests, by quickly dashing into the surf after each wave, grabbing the adults by hand and shoveling small scoops of sand into which their eggs presumably had been laid. These "nests" were then taken back to the lab and incubated in seawater for more than a week, until the embryos hatched.

Trapping. Our most common method for collecting small rodents has entailed the use of metal live traps baited with peanut butter or oatmeal. On one memorable 2-week collecting trip through deserts of the southwestern United States and northwestern Mexico, JCA and three of his colleagues set out 1000 live traps per night, in trap lines radiating out from the base of each hillside. Each metal trap is a small box about $2'' \times 2'' \times 8''$. When a mouse enters the trap to feed, it trips a lever such that a door closes behind it. In the early morning of the next day, we picked up the traps (remembering exactly where each had been placed in the rocky terrain) and examined them for occupants. A good night might yield a 5% success rate (i.e., about 50 mice), whereas a poor night might see us capture only 10–15 animals of several rodent taxa (some of which would not be the targeted species of *Peromyscus*).

We have also found baited traps to be useful in capturing small freshwater turtles, such as mud and musk turtles. In this case, each 2-foot-long trap was made of wire mesh and had funnel-shaped ends that an unsuspecting turtle could enter but not easily exit. The bait was an open can of sardines or perhaps a chicken neck from a local grocery store. Multiple traps would be set out overnight in shallow waters of a promising swamp or marsh.

Digging. One species of native mouse (*Peromyscus polionotus*) required a different collecting protocol. This species lives in tubular burrows that the mice have dug into sandy soils (e.g., along the shoulder of a road). This burrow, the narrow entrance of which is marked by a characteristic mound, slants downward for several feet before opening into a nest cavity about 8 inches in diameter, wherein may reside several members of a mouse family. Leading back from the nest are one or two escape tubes that extend upward to within about an inch of the soil surface. With manual labor, a shovel, and a bit of serendipity, we would capture these animals by digging them out of their burrow system. The resulting "divot" could be as much as 4 feet deep and 6 feet or more long, but the net result in each case was one or a few mice that we caught by hand as they scurried away after exiting their escape tubes.

Another fossorial animal that required considerable digging on our part was the southeastern pocket gopher. Our task entailed excavating a cylindrical hole about 3 feet deep and 4 feet in diameter, centered around one of the many gopher mounds that might be evident in a farmer's field. If dug properly, the walls of our hole would intersect the gopher's underground network of tunnels at several points, into each of which we inserted a spring-operated live trap made from PVC pipe. After waiting for several hours, we would return to see whether a pocket gopher had ambled into one of our traps. Farmers were pleased with our success because they considered gophers to be a serious nuisance.

Hunting. Sometimes we have collaborated with licensed hunters to obtain our specimens. For example, we joined with federal or state wildlife officers at check-in stations for duck hunters in Texas and likewise at check-in stations for deer hunters in South Carolina. In each case, we took small pieces of tissue (such as from liver) from each animal carcass as it was being processed and stored the sample in suitable buffer solution (or perhaps frozen in dry ice or liquid nitrogen) until our return to the laboratory.

Netting. Various kinds of hand-held nets were used to collect a diversity of taxa. The species collected in this fashion ranged from mosquitofish dipped from the margins of small ponds to fruit flies captured using butterfly nets swept above buckets filled with smashed-banana bait. Perhaps the oddest use of hand nets involved our running capture of armadillos by a large hoop net attached to a 1.5-meter pole.

Scavenging. For several years, tower-killed birds were a rich but sad source of avian carcasses for our genetic studies. The 1100-foot-tall T.V. tower that inadvertently knocked down hundreds of birds during migration happened to be located immediately adjacent to the Tall Timbers Research station in northern Florida. Ecologists on the Tall Timbers staff routinely collected avian corpses from around the base of the tower and stored them in their freezers for various scientific research projects, including our own.

Gathering. Many species used in our research were collected from their natural habitats simply by hand. This was especially true of many invertebrate animals. For example, pregnant crayfish were captured from streams, whelk snails and their egg cases from intertidal mudflats, sea spiders from rocky intertidal pools, oysters from suitable shorelines, and horseshoe crabs from shallow estuaries.

Diving. All of our histocompatibility work on marine corals and sponges were conducted *in situ* on reefs in the Caribbean. Using SCUBA at 15- to 80-foot depths, we grafted branches together and later scored the bioassays for the acceptance versus rejection responses that signaled genetic identity or genetic nonidentity, respectively, of the colonies.

Other methods. We have employed many other collecting procedures over the years, each idiosyncratic to the particular creatures being examined. To pick just one peculiar example, in several of our studies of marine turtles we would follow a female as she hauled herself ashore at night to lay her clutch of 100+ eggs in a pit that she digs in the sand. Being extremely careful not to disturb her nesting effort, we would slink up and snatch one egg from each nest for genetic analysis. Of course, in this and all of our other genetic analyses of various creatures, we also had to go through the oft-lengthy process of obtaining all the necessary collecting permits from the relevant local, state, federal, or international jurisdictions.

Sunfishes (Centrarchidae)

INTRODUCTION

JCA's baccalaureate degree (from the University of Michigan's School of Natural Resources in 1970) was in Fisheries Biology and Management. Thus, it should come as no surprise that much of his subsequent research in molecular ecology and evolution over the ensuing decades has been on fishes. His Master's degree (from the University of Texas in 1971) involved an examination of genetic variability in Mexican cavefishes (see Chapter 4), and his dissertation work for Ph.D. (from the University of California at Davis in 1975) entailed a comparison of molecular evolutionary rates in rapidly speciating minnows (Cyprinidae) versus slower speciating sunfishes (Centrarchidae). The North American sunfishes have long been of special research interest in the JCA laboratory, because of their diverse nature (with about 25 species), their usual abundance in native waterways, their interesting ecologies that include nest-building habits and large clutches that almost beg for molecular appraisals of genetic parentage, and their natural proclivity to hybridize both in natural and contrived settings. The abstracts in this chapter reflect JCA's long-standing fascination with sunfishes, some of his favorite creatures. Several of the earliest papers in this series were conducted while JCA was a laboratory technician at the Savannah River Ecology Laboratory in South Carolina, from 1971 to 1973. But his special interest in the beautiful centrarchid species has persisted to the present day.

Biochemical genetics of sunfish I. Geographic variation and subspecific intergradation in the bluegill, *Lepomis macrochirus*

Avise, J.C. and M.H. Smith. 1974. Biochemical genetics of sunfish I. Geographic variation and subspecific intergradation in the bluegill, Lepomis macrochirus. *Evolution 28:42−56.*

ANECDOTE OR BACKDROP

The bluegill sunfish is probably the most abundant and widespread freshwater fish native to the eastern United States, being found in bodies of water ranging from small rivers to huge reservoirs.

J.C. Avise: Sketches of Nature.
http://dx.doi.org/10.1016/B978-0-12-801945-0.00002-4

Because it is so common, it proved to be a suitable candidate for one of the first range-wide genetic surveys of any fish species in the protein-electrophoretic era. On one extended road trip in February 1972, we collected this species from the Carolinas to Texas, sometimes seining the fish from icy waters that required our use of wetsuits.

ABSTRACT

Electrophoretic variation in proteins encoded by 15 genetic loci was analyzed in 2415 bluegill (*Lepomis macrochirus*) representing 47 populations from 7 southern states. Populations from the Florida peninsula and southeastern Georgia (*L.m. purpurescens*) differ in allelic composition at several loci from populations in central and western Georgia west to Texas (*L.m. macrochirus*), yielding coefficients of genetic similarity below the range typifying continuously distributed conspecific populations in other vertebrates, but quite comparable to previous reports for various semispecies pairs. Within several river drainages in South Carolina and Georgia, bluegill populations are segregating for alleles from both subspecies. A closer examination of genotypic classes in a large population from the intergrade zone confirms that the two subspecies are backcrossing and are apparently fully interfertile. The high correlation of allele frequencies across locales in the hybrid zone is compatible with the hypothesis that the alleles are behaving as neutral markers of intergradation. The pattern of introgression evidences a secondary meeting of allopatrically evolved races. Since populations of *L.m. purpurescens* are largely confined to the Florida peninsula, it is likely that Pleistocene rises in sea level were important in their original isolation from *L.m. macrochirus*. Populations of bluegill within reservoirs are generally homogeneous for frequencies of common alleles at polymorphic loci, but there is significant heterogeneity in allele frequencies between reservoirs within a drainage. The magnitude of this variance is greatest in the intergrade populations of the Savannah River basin and is far less in populations of *L.m. macrochirus*. The bluegill examined may be characterized by three areas of relative genetic uniformity: Florida populations of *L.m. purpurescens*, intergrade populations, and populations of *L.m. macrochirus* to the west.

ADDENDUM

The bluegill is a popular "panfish" that is widely bred commercially and used to stock farm ponds. For this reason, we confined our collecting to more natural or open habitats that probably had not been recently stocked with captive-bred specimens. This means that most of the fish surveyed in this study were captured from rivers or large open-water reservoirs, where reproduction presumably occurs naturally rather than in fish hatcheries.

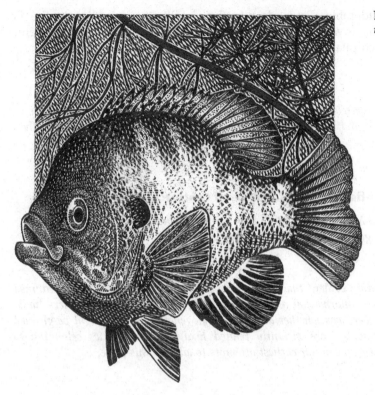

FIGURE 2.1 Bluegill, *Lepomis macrochirus.*

Biochemical genetics of sunfish II. Genic similarity between hybridizing species

Avise, J.C. and M.H. Smith. 1974. Biochemical genetics of sunfish II. Genic similarity between hybridizing species. **American Naturalist** *108:458–472.*

ANECDOTE OR BACKDROP

This study was conducted at a time when there was great interest but little empirical information about the magnitude of genetic divergence between closely related animal species. Early allozyme studies with fruit flies had shown that even sister species might differ in allelic composition at 20% or more of their structural loci, but would such observations extend to other taxa as well? Here we examined genetic divergence among nearly a dozen sunfish species that were presumed to be close evolutionary relatives, based on their well-known propensity to produce interspecific hybrid offspring.

ABSTRACT

Electrophoretic variation in proteins encoded by 18 genetic loci was examined in 1099 specimens of 10 species of sunfish (*Lepomis*), a group renowned for its ability to hybridize both in the laboratory and in nature. Populations belonging to the same species show very little genic differentiation over considerable geographic ranges. Species pairs showed major differences in allelic composition at an average of about one-half of their loci, although the range is considerable. Thus, despite their ability to hybridize to the F_1 generation, *Lepomis* species have very different gene pools. This indicates that close genic similarity is not a necessary corollary of hybridizing propensity in these fishes.

ADDENDUM

It is now known that different vertebrate groups lose their capacity to produce viable hybrids at very different evolutionary rates, and that some vertebrates may retain this capacity for more than 10 million years.

Biochemical genetics of sunfish IV. Relationships of centrarchid genera

Avise, J.C., D.O. Straney, and M.H. Smith. 1977. Biochemical genetics of sunfish IV. Relationships of centrarchid genera. **Copeia 1977:250−258.**

ANECDOTE OR BACKDROP

By the mid 1970s, it was becoming evident that multi-locus protein-electrophoretic data contained at least some useful evolutionary information about conspecific populations and congeneric taxa. What remained unknown, however, was whether any such phylogenetic insights might be gleaned from allozymic comparisons among more distantly related taxa, such as species belonging to confamilial genera. This paper was one of our earliest attempts to address this issue.

ABSTRACT

We examined electrophoretic variation in proteins encoded by 11−14 loci in species representing all 9 genera of Centrarchidae. A dendrogram based on allozyme information is compared to postulated relationships of sunfish genera based on general and specific morphologies, and on hybridizing propensity. The allozyme information correlates most strongly with that derived from a detailed study of the acoustico-lateralis system by Branson and Moore. Similarities between the two sets of data are observed (i) in the clustering together of species of *Lepomis*, (ii) in the clustering of *Lepomis* with *Micropterus*, (iii) in the placement of *Acantharchus* with *Archoplites*, and (iv) in the very distant relationship of *Elassoma* to the other centrarchid genera. Levels of genetic similarity (S) between centrarchid genera are compared to previously published levels of similarity between congeneric species of *Lepomis*, subspecies of *Lepomis macrochirus*, and geographic populations within the subspecies *L.m. macrochirus* and *L.m. purpurescens*. Mean values of S are as follows: between genera, $S = 0.29$; between congeneric species, $S = 0.53$; between subspecies of *Lepomis macrochirus*, $S = 0.85$; and between geographic populations, $S = 0.97$.

ADDENDUM

Systematists later moved one of these centrarchid genera—Elassoma or pygmy sunfishes—to a different taxonomic family, the Elassomatidae.

Gene frequency comparisons between sunfish (Centrarchidae) populations at various stages of evolutionary divergence

Avise, J.C. and M.H. Smith. 1977. Gene frequency comparisons between sunfish (Centrarchidae) populations at various stages of evolutionary divergence. **Systematic Zoology** *26:319–335.*

ANECDOTE OR BACKDROP

This paper summarized our findings to that date about the magnitude of genetic (protein-electrophoretic) divergence between fish populations at various hierarchical stages of evolutionary and taxonomic differentiation. The data were analogous (and proved to be similar) to those that previously had been published by other researchers for populations of fruit flies at comparable stages of evolutionary separation.

ABSTRACT

A total of 11–15 genetic loci encoding enzymes and other proteins were assayed in populations belonging to 20 taxa of North American sunfish in the family Centrarchidae. Four stages of evolutionary divergence are recognized among these taxa by systematists. We have determined mean levels of genetic differentiation (D) between populations at each of these stages: geographic populations, $D = 0.024$; subspecies, $D = 0.171$; species, $D = 0.626$; and genera, $D = 1.340$. The metric D estimates the number of electrophoretically detectable codon substitutions per locus accumulated since populations separated from a common ancestor. Mean genetic distance among centrarchids increases dramatically through the various stages of evolutionary divergence. Different subspecies of *Lepomis macrochirus* exhibit about an eightfold increase in genetic distance over geographic populations belonging to a subspecies. Species of the genus *Lepomis* are completely distinct in allelic composition at nearly 50% of their loci, on the average, despite their ability to produce viable and sometimes fertile hybrids. Species belonging to different centrarchid genera have accumulated more than one allelic substitution per locus, on average. Among the centrarchids, prezygotic isolating barriers may have preceded the development of significant postzygotic isolating mechanisms. Nonetheless, mean levels of genetic divergence between centrarchid taxa are similar to those for other animal taxa of corresponding rank characterized by different modes of reproductive isolation. Amounts and patterns of genetic divergence are employed to infer probable evolutionary relationships among the Centrarchidae. A great deal of evolutionary information is contained within even a relatively small number of genes.

Is evolution gradual or rectangular? Evidence from living fishes

Avise, J.C. 1977. Is evolution gradual or rectangular? Evidence from living fishes. **Proceedings of the National Academy of Sciences USA** *74:5083–5087.*

ANECDOTE OR BACKDROP

In 1972, Niles Eldredge and Stephen J. Gould published an influential but controversial article claiming that evolution proceeds not at a steady pace (phyletic gradualism) but rather via phenotypic jumps (punctuated equilibria) each associated with a speciation event. Their evidence came from a close inspection of fossil records for various taxa, but JCA recognized that their evolutionary model could also have profound implications for the expected mean magnitudes of evolutionary divergence among extant taxa. In particular, extant members of a rapidly speciating clade should have accumulated much more differentiation than extant members of a slowly speciating clade of comparable age, because they would have had many more speciation events in their evolutionary histories. This paper provides an early empirical test of the competing predictions of phyletic gradualism versus punctuated equilibrium, based on observed genetic distances between extant species of two families of North American fishes with very different speciation rates.

ABSTRACT

The traditional view that most evolutionary change is gradual and cumulative within lineages (phyletic gradualism) has recently been challenged by the proposition that the majority of evolutionary change is concentrated within speciation events (rectangular evolution). The logical implications of these competing hypotheses for the means and variances of genetic distance among living members of rapidly and slowly speciating phylads are examined. An example of a critical test of gradual versus rectangular evolution is provided by electrophoretic analyses of genic composition in 69 species of North American Cyprinidae (minnows) and 19 species of Centrarchidae (sunfish). Rate of protein evolution appears somewhat decelerated, if anything, in the rapidly speciating minnows. Results are inconsistent with predictions of rectangular evolution, but are not demonstrably incongruent with predictions of phyletic gradualism.

ADDENDUM

The broader debate between punctuated equilibrium and phyletic gradualism reached its apogee in the 1970s, but has since noticeably faded from discussion in much of the evolutionary literature.

Genic heterozygosity and rate of speciation

Avise, J.C. 1977. Genic heterozygosity and rate of speciation. **Paleobiology 3:422–432.**

ANECDOTE OR BACKDROP

Fisher's fundamental theorem of population genetics predicts that evolutionary rates within populations are proportional to the magnitudes of additive genetic variation (heterozygosity) within those populations. Does this imply that speciation rates also are proportional to heterozygosity levels? In particular, might low levels of heterozygosity in effect put brakes on speciational processes? This paper addresses such issues, based on an empirical comparison of mean allozyme heterozygosities in minnows versus sunfishes.

ABSTRACT

The hypothesis is proposed that mean level of heterozygosity is functionally related to rate of speciation in evolutionary phylads. Under this hypothesis, phylads that speciate more rapidly do so because of increased levels of within-species genetic variability which are then available for conversion to species differences under appropriate ecological or environmental conditions. An important corollary is that rate of speciation could be limited in phylads with low genetic variability, irrespective of environmental considerations. This hypothesis has herein been tested with respect to electrophoretically detectable variation in products of structural genes in two families of North American fishes characterized by grossly different rates of speciation. Totals of 69 species of the highly speciose Cyprinidae, and 19 species of the relatively depauperate Centrarchidae, were assayed for mean level of population heterozygosity (H) at 11–24 genetic loci. Since Cyprinidae and Centrarchidae proved to exhibit on the average nearly identical levels of genic variation (mean $H = 0.052$ and mean $H = 0.049$, respectively), the hypothesis that level of heterozygosity affects rate of speciation in these fishes is not supported. Nonetheless, the amount of genetic variability in both the Cyprinidae and the Centrarchidae is large, comparable to mean levels in previously studied vertebrates. The great wealth of genome variability, reflected in the electrophoretic variation present in virtually all outcrossing organisms, apparently can accommodate considerable flexibility in rate and pattern of evolutionary response to the various environmental regimes that challenge organisms.

Population structure of freshwater fishes. I. Genetic variation of bluegill (*Lepomis macrochirus*) populations in man-made reservoirs

Avise, J.C. and J. Felley. 1979. Population structure of freshwater fishes. I. Genetic variation of bluegill (Lepomis macrochirus) *populations in man-made reservoirs.* Evolution 33:15–26.

ANECDOTE OR BACKDROP

It happens that each of two parallel river drainages in South Carolina (the Santee-Cooper) and Georgia (the Savannah River) has four major reservoirs created by huge artificial dams. It also happens that bluegill sunfish are abundant in all of these bodies of water. The intent of this study was to take advantage of a nested hierarchical sampling design to examine how genetic variation is partitioned within and among bluegill populations from these two drainage systems. The major logistical challenge was collecting 40 fish from each of 64 sites (8 sites equally spaced around each of the 8 reservoirs). To do this required lots of seining from public-access sites such as boat-launching ramps that we located using detailed topographic maps for each of the eight reservoirs.

ABSTRACT

Population structure in the freshwater sunfish *Lepomis macrochirus* was analyzed by examining genotypes at three polymorphic loci in 2560 individuals representing 64 localities distributed evenly among 8 reservoirs and 2 drainages in the southeastern United States. There is no evidence of inbreeding within localities, but allele frequencies among localities within a reservoir are often heterogeneous; mean standardized variance (F_{ST}) for localities of a reservoir was 0.029, consistent across loci. This magnitude of differentiation is slight, far less than for snail populations within cities or house mice on adjacent farms, and occurs despite the immense sizes of the reservoirs sampled (up to 100,000 acres and 1200 shoreline miles). It seems realistic to regard a reservoir population as a quasi-panmictic assemblage of local populations between which differentiation is minimal and only weakly oriented to distance of geographic separation. Variances in allele frequencies among reservoirs within a drainage are much greater ($F_{ST} = 0.392$), accounting for nearly 90% of the total variance observed in the study. Adjacent reservoirs are more similar in allele frequency than are those farther apart. A roughly parallel allelic cline among reservoirs of the two drainages may reflect preexisting heterogeneity among the founding populations in the linear river habitat. Allele frequencies in these reservoirs have not changed significantly during 5 years of observation. The adaptive significance, if any, of the alternative allelic states is unknown. Furthermore, contemporaneous processes affecting population structure, whether deterministic or stochastic, must act upon a preexisting structure derived from a unique set of historical developments that are seldom known for any species. Thus, by hard criteria, definitive statements about the causal processes responsible for population structure are rarely warranted, even in case studies such as the bluegill where the empirical results appear relatively straightforward.

Population structure of freshwater fishes. II. Genetic and morphological variation of bluegill populations in Florida lakes

Felley, J.D. and J.C. Avise. 1980. Population structure of freshwater fishes. II. Genetic and morphological variation of bluegill populations in Florida lakes. **Transactions of the American Fisheries Society** *109:108–115.*

ANECDOTE OR BACKDROP

Having established how genetic variation of bluegill populations is apportioned within and among man-made reservoirs in the southeastern United States (see the previous abstract), we were anxious to learn whether genetic variation might be arranged differently in natural lakes. To do so required that we travel to Florida, where bluegills are likewise abundant and easily collected by seining.

ABSTRACT

The structure of bluegill (*Lepomis macrochirus*) populations in natural Florida lakes was assessed from electrophoretic and morphologic characters. Electrophoretically assayed allele frequencies are homogeneous within lakes and heterogeneous among lakes. Populations in two adjacent lakes connected by a short river are nearly identical. Several considerations indicate that even young-of-the-year bluegills are well mixed within subpopulations of these lakes, and that individual dispersal is important in maintaining inter-subpopulation homogeneity in allele frequency. This pattern of genetic differentiation contrasts with the pattern of heterogeneity observed in meristic counts for several morphological traits. At the microgeographic level, genetic homogeneity probably reflects a long-term history of bluegill movements within a lake, while within-lake morphological heterogeneity reflects the varied conditions during individual development to which incompletely isolated subpopulations are exposed.

Speciation rates and morphological divergence in fishes: tests of gradual versus rectangular modes of evolutionary change

Douglas, M.E. and J.C. Avise. 1982. Speciation rates and morphological divergence in fishes: tests of gradual versus rectangular modes of evolutionary change. Evolution 36:224–232.

ANECDOTE OR BACKDROP

Michael Douglas was my first graduate student at the University of Georgia, and he joined my lab at a time when I had not yet received any external grant funding. This meant that we had to devise a dissertation project that required basically no monetary support. So, using only a $50 set of dial calipers and lots of elbow grease, Mike measured and compared umpteen morphological characters in large numbers of museum specimens (preserved in alcohol) of many sunfish and minnow species. Mike analyzed the resulting volumes of data using a wide variety of multivariate statistical approaches and interpreted all of the findings in the context of the ongoing debate between phyletic gradualism and punctuated equilibrium. Although the entire project entailed no genetic analyses (and ergo no laboratory money), it proved to be especially germane for testing the original punctuated equilibrium proposal (which explicitly stressed the topic of phenotypic evolution per se, rather than molecular genetic evolution).

ABSTRACT

Two competing hypotheses to account for the evolutionary diversification of life are considered. "Phyletic gradualism" predicts that major morphological changes are attributable to the cumulative effects through time of microevolutionary forces (e.g., mutation, drift, and selection), which provide gradual divergence among reproductively isolated lineages. "Punctuated equilibria," or "rectangular evolution," predicts that magnitude of macroevolutionary divergence is governed primarily by cladogenetic speciation, since most morphological change occurs only during the process of speciation itself. Critical tests of the original rectangular evolution/phyletic gradualism models are provided in this report, based on a variety of multivariate analyses of 50 morphometric characters measured in each of 11 species of *Lepomis* (sunfish) and in 37 of the 100+ species of *Notropis* (minnows). Rate of morphological divergence in the rapidly speciating *Notropis* phylad is no greater than rate of morphological change in the relatively slow-speciating *Lepomis*. Results are inconsistent with some of the predictions of rectangular evolution.

ADDENDUM

This study remains one of the few morphology-based appraisals of punctuated equilibrium versus phyletic gradualism that expressly uses neontological data from extant species representing rapidly versus slowly speciating clades.

Characterization of mitochondrial DNA variability in a hybrid swarm between subspecies of bluegill sunfish (*Lepomis macrochirus*)

Avise, J.C., E. Bermingham, L.G. Kessler, and N.C. Saunders. 1984. Characterization of mitochondrial DNA variability in a hybrid swarm between subspecies of bluegill sunfish (Lepomis macrochirus). Evolution 38:931–941.

ANECDOTE OR BACKDROP

This study was conducted when my laboratory had just recently begun to assay mitochondrial (mt) DNA in rodents (see Chapter 13). At that time, virtually nothing was known about the molecular composition, transmission genetics, or evolutionary patterns of mtDNA in any creatures other than fruit flies and humans or other mammals. Thus this study was highly exploratory and novel at several levels: in addressing molecular variation of mtDNA in a fish; in studying the evolution and transmission genetics of mtDNA in a hybrid-zone setting; and in exploring geographic variation in mtDNA in what later would be called a phylogeographic context.

ABSTRACT

We begin to characterize the evolutionary dynamics of mitochondrial (mt) DNA in fishes by examining restriction site variability in 189 bluegill sunfish (*Lepomis macrochirus*). Fifteen endonucleases were employed to map 37 restriction sites from selected individuals. Genome size was approximately 16.2 kilobases. All differences between genotypes could be accounted for by gains or losses of individual restriction sites, without additions, deletions, or rearrangements affecting more than about 50–250 base pairs. Two highly distinct mtDNA genomes, differing by 20 assayed mutation steps and an estimated 8.5% sequence divergence, were discovered. Both genomes were observed in high frequency in a sample of 151 bluegill from a north Georgia population that on the basis of allozyme genotype appears to represent a freely interbreeding hybrid swarm between two bluegill subspecies. Within the hybrid population, mtDNA and allozyme genotypes were associated approximately at random. The distinct genetic markers provided by mtDNA provided an important test of the possibility of within-individual mtDNA polymorphism (because true heteroplasmy could be distinguished from results of incomplete digests). Nonetheless, in somatic or germ cells, no bluegill exhibited mtDNA heteroplasmy (at a 5% or greater level) that could be explained by paternal mtDNA transmission. Additional samples from Louisiana to Florida tentatively confirm that the geographic distributions of the two distinct mtDNA genomes are highly concordant with the previously described ranges of *L.m macrochirus* and *L.m. purpurescens* as defined by morphology and allozymes. Overall, results on mtDNA variability in bluegill conform to and strengthen some of the straightforward expectations about the pattern of evolution of uniparentally transmitted genomes in sexually reproducing populations.

ADDENDUM

Today this analysis might be called a "cytonuclear dissection" (see Chapters 3 and 8) of a hybrid zone.

Hybridization and introgression among species of sunfish (*Lepomis*): analysis by mitochondrial DNA and allozyme markers

Avise, J.C. and N.C. Saunders. 1984. Hybridization and introgression among species of sunfish (Lepomis): analysis by mitochondrial DNA and allozyme markers. **Genetics 108:237–255.**

ANECDOTE OR BACKDROP

This study was designed to explore the previously untapped potential of mtDNA—especially in conjunction with nuclear genetic data—to genetically dissect the phenomenon of interspecific hybridization, again using sunfish as a prototypic taxonomic group. I randomly collected (by hook-and-line) several hundred specimens of Lepomis from rivers and streams near my home-base of Athens, GA, carefully checking each one to see whether it might be an interspecific hybrid based on morphology. Each fish then had to be hurriedly returned to the laboratory, because in those early years (long before PCR) we had to use fresh liver tissue for our mtDNA isolations using CsCl gradient centrifugation. To our surprise and delight, mtDNA in fishes proved to display all of the favorable molecular and populational properties that we had likewise been uncovering in mammals. And, as this study showed, mtDNA could be hugely informative in helping us to genetically dissect behavioral and evolutionary phenomena at work in hybrid settings.

ABSTRACT

We explore the potential of mitochondrial (mt) DNA analysis, alone and in conjunction with allozymes, to study low-frequency hybridization and introgression phenomena in natural populations. MtDNAs from small samples of 9 species of sunfish (*Lepomis*) were purified and digested with each of 13 informative restriction enzymes. Digestion profiles for all species were highly distinct: estimates of overall fragment homology between pairs of species ranged from 0% to 36%. Allozymes encoded by nine nuclear genes also showed large frequency differences among species and together with mtDNA provided many genetic markers for hybrid identification. A genetic analysis of 277 sunfish from two locations in north Georgia revealed the following: (i) a low frequency of interspecific hybrids, all of which appeared to be F_1s; (ii) the involvement of five sympatric *Lepomis* species in the production of these hybrids; (iii) no evidence for introgression between species in our study locales (although for rare hybridization, most later-generation backcrosses would not be reliably distinguished from parentals); (iv) a tendency for hybridizations to take place preferentially between parental species differing greatly in abundance; and (v) a tendency for rare species in a hybrid cross to provide the female parent. Our data suggest that an absence of conspecific pairing partners and mating stimuli for females of rare species may be important factors in increasing the likelihood of interspecific hybridization. The maternal inheritance of mtDNA offers at least two novel advantages for hybridization analysis: (i) an opportunity to determine direction in hybrid crosses and (ii) due to the linkage among mtDNA markers, an increased potential to distinguish effects of introgression from either symplesiomorphy or character convergence.

Cytonuclear introgressive swamping and species turnover of bass after an introduction

Avise, J.C., P.C. Pierce, M.J. Van Den Avyle, M.H. Smith, W.S. Nelson, and M.A. Asmussen. 1997. Cytonuclear introgressive swamping and species turnover of bass after an introduction. Journal of Heredity *88:14–20.*

ANECDOTE OR BACKDROP

This project was initiated when Georgia's Department of Natural Resources (DNR) approached us to genetically analyze a peculiar situation in a northern Georgia lake. Over a period of several years, following an unauthorized release of Spotted Bass into Lake Chatuge, native smallmouth bass had apparently become quite rare. Using tissue samples from a large number of adult bass that DNR personnel collected by electroshocking, we were able to genetically document a remarkable example of "introgressive swamping," wherein the gene pool of a population has shifted dramatically, in part as a result of extensive interspecific hybridization. Indeed, based on our cyto-nuclear genetic analyses, in Lake Chatuge there had been a nearly complete turnover or replacement of one species' genes with those of a closely related species.

ABSTRACT

Species-specific RFLP markers from mitochondrial (mt) DNA were identified and employed in conjunction with previously reported data from nuclear allozyme markers to examine the genetic consequences of an artificial introduction of spotted bass (*Micropterus punctulatus*) into a north Georgia reservoir originally occupied by native smallmouth bass (*M. dolomieui*). The cytonuclear genetic data indicate that within 10–15 years following the unauthorized introduction, a reversal in these species' abundances has occurred and that more than 95% of the population sample now consists of spotted bass or products of interspecific hybridization. This demographic shift, perhaps ecologically or environmentally mediated, has been accompanied by introgressive swamping; more than 95% of the remaining smallmouth bass nuclear and cytoplasmic alleles are present in individuals of hybrid ancestry. Dilocus cytonuclear disequilibria were significantly different from zero, with patterns indicative of an excess of homospecific genetic combinations (relative to expectations from single locus allelic frequencies) and a disproportionate contribution of smallmouth bass mothers to the hybrid gene pool. Results document that dramatic genetic and demographic changes can follow the human-mediated introduction of a non-native species.

ADDENDUM

This was an early study that helped to put the phenomenon of introgressive swamping (as another possible source of population extinction) on the radar screen of conservation biologists.

FIGURE   2.2 Spotted   Bass, *Micropterus punctulatus.*

Molecular genetic dissection of spawning, parentage, and reproductive tactics in a population of redbreast sunfish, *Lepomis auritus*

DeWoody, J.A., D.E. Fletcher, S.D. Wilkins, W.S. Nelson, and J.C. Avise. 1998. Molecular genetic dissection of spawning, parentage, and reproductive tactics in a population of redbreast sunfish, Lepomis auritus. *Evolution 52:1802−1810.*

ANECDOTE OR BACKDROP

In all sunfish (centrarchid) species, at least some of the adult males build and defend saucer shaped nests into which one or more females may lay eggs. These nests offer great opportunities for assessing both paternal and maternal contributions to a brood, based on genetic parentage analyses of the embryos within each nest. The main laboratory challenge is to find suitably polymorphic markers, which batteries of microsatellite loci often provide. The main field challenge is to collect earmarked nests and their attendant males from nature, a task that we accomplished by electroshocking each attendant male and then scooping up his nest and its contents. This study was the first of several such analyses of genetic maternity and genetic paternity from the large collection of embryos within each nest of centrarchid and various other nest-tending fish species (see also Chapters 4, 6, and 16).

ABSTRACT

We developed and employed microsatellite markers to describe genetic paternity and maternity of progeny cohorts in a population of redbreast sunfish (*Lepomis auritus*), a species in which males build and tend nests. A total of nearly 1000 progeny from 25 nests, plus nest-attendant males and nearby females, were genotyped at microsatellite loci that displayed more than 18 alleles each (combined exclusion probabilities greater than 0.90). The genetic data demonstrate that multiple females (at least 2−6) spawned in each nest; that their offspring were spatially dispersed across the nest; and that more than 90% of the fry in most nests were sired by their attendant male. However, about 40% of the nests also showed genetic evidence of low-level reproductive parasitism, and two nests were tended by males who had fathered none of the fry. Overall, the data indicate a genetically promiscuous spawning system for this population, with paternity attributed predominantly to the nest-attendant males. This pattern contrasts with a reproductive mode reported previously in bluegill sunfish (*L. macrochirus*) wherein heteromorphic males specialized for parasitism or for parental care co-occur in high frequencies.

ADDENDUM

For the bluegill sunfish, other researchers similarly had used molecular markers to document that fry within a nest could be sired by any of three types of males: "bourgeois" males that build and tend the nests; sneaker males that slip onto a nest to release sperm during a spawning episode; and satellite males that mimic females in appearance and behavior and thereby also gain access to a nest for spawning.

FIGURE 2.3   Redbreast sunfish, *Lepomis auritus.*

The genetic mating system of spotted sunfish (*Lepomis punctatus*): mate numbers and the influence of male reproductive parasites

*DeWoody, J.A., D.E. Fletcher, M. Mackiewicz, S.D. Wilkins, and J.C. Avise. 2000. The genetic mating system of spotted sunfish (*Lepomis punctatus*): mate numbers and the influence of male reproductive parasites. Molecular Ecology 9:2119–2128.*

ANECDOTE OR BACKDROP

In addition to the nest-tending or "bourgeois" males, some centrarchid and other fish species have parasitic cuckolder males that may "steal" some of the fertilizations from the nestholders. Such sneaker males may also show behavioral and morphological adaptations for the parasitic lifestyle, such as a much greater relative investment in gonadal as opposed to somatic tissues. This study presents a genetic analysis of paternity (and maternity) in one such species with a low frequency of males specialized for nest parasitism.

ABSTRACT

In nest-building fish species, mature males often exhibit one of two alternative reproductive behaviors. Bourgeois males build nests, court females, and guard their eggs. Parasitic cuckolders attempt to steal fertilizations from bourgeois males and do not invest in parental care. Previous evidence from the bluegill sunfish (*Lepomis macrochirus*) suggests that adult males are morphologically specialized for these two tactics. Here, we used microsatellite markers to determine genetic parentage in a natural population of the spotted sunfish (*L. punctatus*) that also displayed both bourgeois and parasitic male morphs. As gauged by relative investments in gonadal versus somatic tissues, between 5% and 15% of the mature adult males were parasites. Multi-locus genotypes were generated for more than 1,400 embryos in 30 nests, their nest-guardian males, and for other adults in the population. Progeny in about 57% of the nests were sired exclusively by the guardian male, but the remaining nests contained embryos resulting from cuckoldry as well. Overall, the frequency of offspring resulting from stolen fertilizations was only 1.3%, indicating that the great majority of paternity is by bourgeois nesting males. With regard to maternity, 87% of the nests had at least 3 dams, and computer simulations estimate that about 7.2 dams spawned per nest.

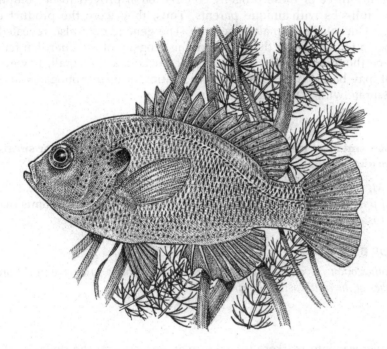

FIGURE 2.4   Spotted sunfish, *Lepomis punctatus.*

Genetic monogamy and biparental care in an externally fertilizing fish, the largemouth bass (*Micropterus salmoides*)

DeWoody, J.A., D.E. Fletcher, S.D. Wilkins, W.S. Nelson, and J.C. Avise. 2000. Genetic monogamy and biparental care in an externally fertilizing fish, the largemouth bass (Micropterus salmoides). Proceedings of the Royal Society of London B 267:2431–2437.

ANECDOTE OR BACKDROP

Polygamy and exclusive male care of offspring are common in nest-tending fishes, whereas monogamy and biparental care are generally thought to be exceedingly rare. This study appears to provide a notable exception to these trends.

ABSTRACT

In North American sunfish (Centrarchidae), breeding males are often brightly colored and promiscuous. However, the largemouth bass (*Micropterus salmoides*) is sexually monomorphic in appearance and socially monogamous. Unlike some other nest-tending centrarchids in the genus *Lepomis*, largemouth bass has been reported to provide biparental care to eggs and fry. Here we use microsatellite markers to ask whether social monogamy predicts genetic monogamy in the largemouth bass. Offspring were collected from 26 nests each usually guarded by a pair of adults, many of which were also captured. Twenty three of these progeny cohorts (88%) proved to be composed almost exclusively of full-sibs with unique parents. Thus, they were the product of monogamous matings. Cuckoldry by males was rare. The genetic data also revealed that some nests contained a few offspring that were not the progeny of the guardian female, a finding that can be thought of as low-level "female cuckoldry." Overall, however, the data provide what may be the first genetic documentation of near-monogamy and biparental care in a vertebrate with external fertilization.

A genetic assessment of parentage in a natural population of dollar sunfish (*Lepomis marginatus*) based on microsatellite markers

*Mackiewicz, M., D.E. Fletcher, S.D. Wilkins, J.A. DeWoody, and J.C. Avise. 2002. A genetic assessment of parentage in a natural population of dollar sunfish (*Lepomis marginatus*) based on microsatellite markers.* **Molecular Ecology 11:1877–1883.**

ANECDOTE OR BACKDROP

This is yet another of our several microsatellite analyses of genetic parentage in the broods of nest-tending sunfishes of the southeastern United States.

ABSTRACT

We employ microsatellite markers to assess mating tactics in the dollar sunfish, *Lepomis marginatus*. Genetic assignments for 1,015 progeny in 23 nests indicate that about 95% of the offspring were sired by their respective nest-guardians, a finding consistent with the apparent absence of a brood parasitic morphotype in this species. Allopaternal care was documented in two nests, one resulting from a nest takeover, the other from cuckoldry by an adjoining nest-tender. Clustered *de novo* mutations also were identified. About 2.5 females (range 1–7) contributed to the offspring pool within a typical nest. Results are compared to those for other *Lepomis* species.

ADDENDUM

The Dollar Sunfish occurs on coastal-plain drainages from North Carolina to Florida and west to Texas. It is a rather small but especially beautiful centrarchid species that spawns in solitary sand nests (usually adjacent to a log or other such structure) from about April to October.

FIGURE 2.5   Dollar Sunfish, *Lepomis marginatus*.

Live-bearing Fishes (Poeciliidae)

INTRODUCTION

Pregnancy and livebearing (viviparity) are not confined to mammals. These phenomena also characterize many other organisms, including freshwater fishes in the New World family Poeciliidae. In these inch-long creatures, impregnated females may carry dozens of offspring, internally, before eventually giving birth to live young. Because a pregnant female is physically associated with the embryos she bears, opportunities arise to use molecular markers from each pregnant mother and her brood to decipher genetic paternity within the clutch as well (much as in conventional paternity analyses in humans). Some of the abstracts in this chapter describe this analytical procedure.

Even more remarkable is the fact that some poeciliid fishes are unisexual species that consist solely of females. These females reproduce by mechanisms related to parthenogenesis (virgin birth) or other similar procreative modes. For several years in the late 1980s and early 1990s, JCA collaborated with Robert Vrijenhoek and others to study the evolutionary genetics of these all-female biotypes in nature. Mitochondrial (mt) DNA proved to be especially useful for these appraisals because the matrilineal history recorded in the mitochondrial genome is in principle one-and-the-same as the entire organismal phylogeny. This outcome stands in striking contrast to the standard situation in sexual taxa, in which the matriarchal phylogeny recorded by mtDNA constitutes only a minuscule fraction of a species' total hereditary history (most of which is ensconced in a nuclear genome that was transmitted across the generations through both sexes). And in an interesting turn of the tables, some unisexual fishes also offer unique insights about the transmission genetics of mtDNA, as described by the first abstract in this collection.

J.C. Avise: Sketches of Nature.
http://dx.doi.org/10.1016/B978-0-12-801945-0.00003-6

Mode of inheritance and variation of mitochondrial DNA in hybridogenetic fishes of the genus *Poeciliopsis*

Avise, J.C. and R.C. Vrijenhoek. 1987. Mode of inheritance and variation of mitochondrial DNA in hybridogenetic fishes of the genus Poeciliopsis. *Molecular Biology and Evolution 4:514–525.*

ANECDOTE OR BACKDROP

By the mid-1980s, evidence had accumulated from several vertebrate and invertebrate species that mitochondrial (mt) DNA is normally transmitted across the generations primarily if not exclusively via females. However, several reports also had hinted at the possibility of low-level or sporadic paternal input of mtDNA ("paternal leakage") into the system. In 1985, JCA teamed with Robert Vrijenhoek to critically test for paternal leakage, using what in effect is a backcross regimen that nature herself provides. As described in this next abstract, that natural detection system involved hybridogenetic Poeciliopsis *fishes from northwestern Mexico, the females of which in effect are engaged in a perpetual backcross to males of a related species. Thus, if even a tiny fraction of mtDNA is paternally derived in each generation, it should quickly build up to detectable levels in later-generation progeny. This study provides a critical test of that possibility.*

ABSTRACT

A genetic survey of mitochondrial (mt) DNA in a hybridogenetic complex of fishes (genus *Poeciliopsis*) was conducted to assess the possibility of low-level paternal transmission of mtDNA to progeny. In this reproductive system, females of the unisexual hybrid biotype between *P. monacha* and *P. lucida* effectively participate in a perpetual backcross to males of a sexual species, *P. lucida*. As judged on the basis of numerous restriction digests, mtDNAs of the bisexual parental species (*P. monacha* and *P. lucida*) were highly distinct, yet mtDNA of the natural hybridogens was not different from that of *P. monacha* from the same river system. Since the hybridogens are probably thousands of generations old, the present results demonstrate that paternal leakage of mtDNA must be extremely low or absent in these fishes. mtDNA genotypic differences among hybridogenetic strains were also present and corresponded to geographic locale. These differences provide a foundation for estimation of both origins and phylogeny of the unisexual forms.

ADDENDUM

For another such refined test for low-level paternal leakage of mtDNA, see in Chapter 15 *the paper entitled "Critical experimental test of the possibility of 'paternal leakage'..."*

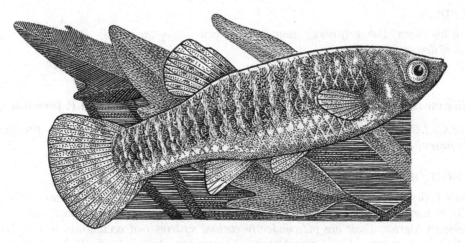

FIGURE 3.1   Headwater Livebearer, *Poeciliopsis monacha-lucida*.

Molecular evidence for multiple origins of hybridogenetic fish clones (Poeciliidae: *Poeciliopsis*)

Quattro, J.M., J.C. Avise, and R.C. Vrijenhoek. 1991. Molecular evidence for multiple origins of hybridogenetic fish clones (Poeciliidae: Poeciliopsis*). Genetics 127:391–398.*

ANECDOTE OR BACKDROP

Bob Vrijenhoek had devoted much of his scientific career to the study of unisexual (all-female) species in the fish genus Poeciliopsis. *In the late 1980s, he and JCA became co-PIs on an NSF grant to analyze these fishes using newly developed techniques for mtDNA analysis. One of Bob's students who became heavily involved with this project was Joe Quattro. In this paper, Joe used mtDNA to show that hybridogenetic lineages of* Poeciliopsis *are highly polyphyletic, having arisen on multiple occasions from crosses involving different females of* P. monacha.

ABSTRACT

Hybrid matings between the sexual species *Poeciliopsis monacha* and *Poeciliopsis lucida* produced a series of diploid all-female lineages of *P. monacha-lucida* that inhabit the Río Fuerte of northwestern Mexico. Restriction site analyses of mitochondrial (mt) DNA clearly revealed that *P. monacha* was the maternal ancestor of these hybrids. The high level of mtDNA diversity in *P. monacha* was mirrored by similarly high levels in *P. monacha-lucida*; thus, hybridizations giving rise to unisexual lineages have occurred many times. However, mtDNA variability among *P. monacha-lucida* lineages revealed a geographical component. Apparently the opportunity for the establishment of unisexual lineages varies among tributaries of the Río Fuerte. We hypothesize that a dynamic complex of sexual and clonal fishes appears to participate in a feedback process that maintains genetic diversity in both the sexual and asexual components.

ADDENDUM

Later in his career, Bob Vrijenhoek went on to become one of the world's leading experts on the genetics of deep-sea organisms.

Poecilia mexicana is the recent female parent of the unisexual fish *P. formosa*

Avise, J.C., J.C. Trexler, J. Travis, and W.S. Nelson. 1991. Poecilia mexicana *is the recent female parent of the unisexual fish* P. formosa. *Evolution 45:1530–1533.*

ANECDOTE OR BACKDROP

Although both groups are in the live-bearing family Poeciliidae, unisexual fishes of the genus Poecilia *in northeastern Mexico are not to be confused with the unisexual* Poeciliopsis *fishes of northwestern Mexico. These are independently evolved systems that occupy different regions and generally have been studied by different teams of researchers. In this study, JCA teamed with Joel Trexler and Joe Travis to elucidate (using mtDNA markers) the direction of cross that gave rise to the unisexual (all-female) biotype* Poecilia formosa. *This was merely the latest in what was to become a long list of unisexual vertebrate taxa for which mtDNA has provided unequivocal documentation of maternal ancestry.*

ABSTRACT

Mitochondrial assays of a gynogenetic complex of fishes in northeastern Mexico reveal that the unisexual fish *Poecilia formosa* arose via hybridization between females representing the sexual species *P. mexicana* and males representing the sexual species *P. latipinna*. This study contributes to the growing catalogue of unisexual vertebrates for which the bisexual female ancestor has now been determined. It also contributes to the emerging view that most unisexual vertebrate species are evolutionarily young, and in terms of matriarchal phylogeny are embedded within the broader matriarchal diversity of their female sexual progenitors.

ADDENDUM

Initially discovered in 1932, the Amazon Molly was the first clonal vertebrate conclusively known to science. Its common name derives from reference to the fictitious all-female tribe of human warriors from the Amazon region. Like several other unisexual taxa that reproduce by various clonal or quasi-clonal mechanisms, this species does indeed consist solely of females.

FIGURE 3.2   Amazon Molly, *P. formosa.*

An ancient clonal lineage in the fish genus *Poeciliopsis* (Atheriniformes: Poeciliidae)

Quattro, J.M., J.C. Avise, and R.C. Vrijenhoek. 1992. An ancient clonal lineage in the fish genus Poeciliopsis *(Atheriniformes: Poeciliidae).* **Proceedings of the National Academy of Sciences USA** *89:348–352.*

ANECDOTE OR BACKDROP

Asexual taxa generally are assumed to be evolutionarily short-lived because they lack the recombinational mechanisms that can repair DNA damages and/or generate the genetic variety that presumably is a prerequisite for adaptive evolution in changing environments. Unisexual taxa offer superb subjects for testing this thesis, assuming that we can estimate their evolutionary ages empirically. Here the authors use mtDNA to document a relatively ancient age (ca. 100,000 generations) for one unisexual lineage of Poeciliopsis *fishes. Although this finding might seem to overthrow conventional wisdom about the ages of asexual taxa, in truth by evolutionary standards 100,000 generations "is but an evening gone." Thus, even though this clonal lineage appears to be rather old, this example alone does not imply that the standard evolutionary dogma about asexual taxa must be overturned.*

ABSTRACT

Genetic diversity in mtDNA was assessed within the unisexual (all-female) hybridogenetic fish *Poeciliopsis monacha-occidentalis* and the two sexual species from which it arose. Results confirm that *P. monacha* was the maternal ancestor and that paternal leakage of *P. occidentalis* mtDNA has not occurred. Of particular interest is the high level of *de novo* mutational divergence within one hybridogenetic lineage that on the basis of independent zoogeographic considerations, protein electrophoretic data, and tissue grafting analysis is of monophyletic (single hybridization) origin. Using a conventional mtDNA clock calibration, we estimate that this unisexual clade might be >100,000 generations old. Contrary to conventional belief, this result shows that some unisexual vertebrate lineages can achieve a substantial evolutionary age.

Mode of origin and sources of genotypic diversity in triploid gynogenetic fish clones (*Poeciliopsis*: Poeciliidae)

Quattro, J.M., J.C. Avise, and R.C. Vrijenhoek. 1992. Mode of origin and sources of genotypic diversity in triploid gynogenetic fish clones (Poeciliopsis: Poeciliidae). **Genetics 130:621–628.**

ANECDOTE OR BACKDROP

To collect specimens for this study, JCA joined Bob Vrijenhoek and Joe Quattro on an exciting field trip to mainland northwestern Mexico. The trip was exhilirating not only for its challenge of seining fish from the local arroyos but also for the fact that we had to be constantly on the lookout for drug smugglers that are known to be common in that wide-open part of the country.

ABSTRACT

Most tributaries of the Río Fuerte in northwestern Mexico contain one or more clones of allotriploid fish of the genus *Poeciliopsis*. We used multilocus allozyme genotypes and mitochondrial (mt) DNA haplotypes to examine several potential modes of origin of these gynogenetic all-female fish. The allozyme studies corroborated earlier morphological work revealing the hybrid constitution of two triploid biotypes, *Poeciliopsis 2 monacha-lucida* and *Poeciliopsis monacha-2 lucida*. Each biotype carries one or two whole genomes from each of the sexual species *P. monacha* and *P. lucida*. Restriction site analysis of mtDNA revealed that *P. monacha* was the maternal ancestor of five electrophoretically distinguishable triploid clones. Four of five clones were marked by closely related, composite, allozyme/ mtDNA genotypes suggesting they had common origins from an allodiploid clone of the *P. monacha-lucida* biotype. Genotypic analysis revealed that all five clones arose via the "genomic addition" pathway. Fertilization of unreduced ova in *P. monacha-lucida* females by sperm from *P. monacha* and *P. lucida* males, respectively, gave rise to both biotypes.

ADDENDUM

Populations of unisexual Poeciliopsis *in northwestern Mexico are in serious decline due to habitat alterations and introductions of exotic fish species. What a shame it would be if these genetically marvelous little fishes were to go extinct.*

Molecular clones within organismal clones: mitochondrial DNA phylogenies and the evolutionary histories of unisexual vertebrates

Avise, J.C., J.M. Quattro, and R.C. Vrijenhoek. 1992. Molecular clones within organismal clones: mitochondrial DNA phylogenies and the evolutionary histories of unisexual vertebrates. Evolutionary Biology 26:225–246.

ANECDOTE OR BACKDROP

This is a review paper that was prompted by our accumulated experience with gynogenetic and hybridogenetic fishes in the genera Poecilia and Poeciliopsis. It occurred to me that a unifying and scientifically original theme would be to ask what special can be learned from the study of a unisexually transmitted genome (mtDNA) in unisexual (all-female) taxa. It turns out that much can be uncovered about the evolutionary origins and subsequent genealogical histories of unisexual parthenogens, gynogens, and hybridogens.

ABSTRACT

Genetic surveys of mitochondrial (mt) DNA diversity have provided a novel class of information on the evolutionary histories of more than 25 "species" of unisexual vertebrates and their bisexual progenitors. A review of this literature reveals that (i) mtDNA inheritance in hybridogenetic fishes is indeed strictly maternal; (ii) most unisexuals arose through nonreciprocal hybridization events between bisexual species in which the female parent has now been identified; (iii) most polyploid unisexuals arose via fertilization of unreduced eggs in a diploid hybrid, rather than spontaneously via fertilization of unreduced eggs in a nonhybrid; (iv) with few exceptions, overall mtDNA genetic diversity within an assayed unisexual taxon is considerably lower than that within its maternal bisexual cognate; (v) in terms of matriarchal phylogeny, a few bisexual–unisexual complexes exhibit a pattern of polyphyly (demonstrating independent hybridization origins of unisexuals from unrelated female ancestors), while most show paraphyly (in which the unisexual appears to have arisen only once or a few times within a subset of the matriarchal genealogy of the sexual parent); (vi) although many unisexuals are closely related to maternal lineages in the sexual relative, and thus appear evolutionarily young, in one well-characterized unisexual clade (involving *Poeciliopsis monacha-occidentalis*), numerous postformational mutations indicate an evolutionary age of at least 100,000 generations. Thus, contrary to conventional wisdom, some unisexual vertebrate clades can achieve considerable evolutionary longevity. Unisexual biotypes provide exceptions to the norm of sexual reproduction in vertebrates; mitochondrial DNA provides an exception to the norm of biparental Mendelian inheritance. These "aberrant" systems studied together have added a synergistic boost to our understanding of the evolutionary genetics of clonal systems.

ADDENDUM

More than a decade later, JCA expanded many of this paper's themes in a book entitled "Clonality: The Genetics, Ecology, and Evolution of Sexual Abstinence in Vertebrate Animals."

Cytonuclear genetic architecture in mosquitofish populations and the possible roles of introgressive hybridization

Scribner, K.T. and J.C. Avise. 1993. Cytonuclear genetic architecture in mosquitofish populations and the possible roles of introgressive hybridization. Molecular Ecology *2:139–149.*

ANECDOTE OR BACKDROP

Mosquitofish are another group of live-bearing fishes that facilitate—by virtue of their abundance and collectability—studies of population genetic and phylogeographic structure of a taxonomic complex of vertebrates across the southeastern United States. The two mosquitofish species in the region also proved to hybridize in a wide zone of secondary contact, and this hybrid zone proved to offer fine fodder for cytonuclear dissections of introgression phenomena. Kim Scribner was a graduate student in JCA's lab who in the early 1990s took the lead role in our genetic analyses of Gambusia.

ABSTRACT

Spatial genetic structure in populations of mosquitofish (*Gambusia*) sampled throughout the southeastern United States was characterized using mitochondrial (mt) DNA and allozyme markers. Both sets of data revealed a pronounced genetic discontinuity (along a broad path extending from southeastern Mississippi to northeastern Georgia) that corresponds to a recently recognized distinction between the nominal forms *G. affinis* to the west and *G. holbrooki* to the east. However, several populations from the general contact region exhibited unusual allelic associations in high frequency, suggestive of evolutionary processes within a zone of introgressive hybridization. These involve: (i) cytonuclear profiles representing combinations of nuclear and mitochondrial genotypes that tended to be more nearly species-specific and concordant elsewhere, and (ii) significant nuclear gametic disequilibria perhaps attributable to positive assortative mating and/or differential fitnesses of homospecific versus recombinant genotypes. However, outside this suspected hybrid region, "heterospecific" genetic markers also appeared in low frequency, thus complicating interpretations. These discordant alleles on a broader geographic scale may reflect: (i) the retention of polymorphisms from an ancestral gene pool, (ii) occasional evolutionary convergence (especially with respect to electrophoretic mobility of allozyme alleles), (iii) the "footprints" of a moving hybrid zone, or (iv) differential introgressive penetrance across the current hybrid region.

ADDENDUM

Mosquitofish get their name from the fact that they like to eat mosquito larvae that live suspended from the surface of a pond or swamp. Indeed, captive-bred mosquitofish are often stocked into bodies of water explicitly for mosquito control.

FIGURE 3.3 Mosquitofish, *Gambusia affinis*.

Population cage experiments with a vertebrate: the temporal demography and cytonuclear genetics of hybridization in *Gambusia* fishes

Scribner, K.T. and J.C. Avise. 1994. Population cage experiments with a vertebrate: the temporal demography and cytonuclear genetics of hybridization in Gambusia *fishes. Evolution 48:155–171.*

ANECDOTE OR BACKDROP

Most of the studies in JCA's lab have involved genetic analyses of populations in the wild. However, on rare occasions opportunities have presented themselves for more experimental approaches under controlled conditions. This study provides one such example, in which Kim Scribner set up—and then temporally monitored—experimental populations of mosquitofish initiated with known numbers of the two species. It greatly surprised us that both of these sets of replicated populations yielded similar cytonuclear trajectories across the 2 years of the study. We both had assumed that nature would be much less predictable or repeatable than that.

ABSTRACT

The dynamics of mitochondrial and multilocus nuclear genotype frequencies were monitored for 2 years in experimental populations established with equal numbers of two poeciliid fishes (*Gambusia affinis* and *Gambusia holbrooki*) that hybridize naturally in the southeastern United States. In replicated "small-pool" populations (experiment I), 1,018 individuals sampled at six time periods revealed an initial flush of hybridization, followed by a rapid decline in frequencies of *G. affinis* nuclear and mitochondrial alleles over 64 weeks. Decay of gametic and cytonuclear disequilibria differed from expectations under random mating as well as under a model of assortative mating involving empirically estimated mating propensities. In 2 replicate "large pond" populations (experiment II), 841 sampled individuals across four reproductive cohorts revealed lower initial frequencies of F_1 hybrids than in experiment I, but again *G. holbrooki* alleles achieved high frequencies over four generations (72 weeks). Thus, evolution within experimental *Gambusia* hybrid populations can be extremely rapid, resulting in a consistent loss of *G. affinis* nuclear and cytoplasmic alleles. Concordance in results between experiments and across genetic markers suggests strong directional selection favoring *G. holbrooki* genotypes. Results are interpreted in light of previous reports of genotype-specific differences in life history traits, reproductive ecology, patterns of recruitment, and size-specific mortality, and in the context of patterns of introgression previously studied indirectly from spatial observations on cytonuclear genotypes in natural *Gambusia* populations.

ADDENDUM

Researchers working with fruit flies often use "population cages" (mesh-encased boxes) to monitor genetic changes in Drosophila populations, but this study on mosquitofish remains one of the few to have used a vertebrate animal for such temporal experimentation.

Cytonuclear genetics of experimental fish hybrid zones inside Biosphere-2

Scribner, K.T. and J.C. Avise. 1994. Cytonuclear genetics of experimental fish hybrid zones inside Biosphere-2. **Proceedings of the National Academy of Sciences USA 91:5066—5069.**

ANECDOTE OR BACKDROP

In the early 1990s, JCA was briefly appointed to a board that oversaw operations of the Biosphere-2 project near Tucson, AZ. Biosphere-2 is a man-made enclosure meant to simulate what life might be like for astronauts seeking to colonize another planet. The Biospherians that inhabited Biosphere-2 for 2 years had to live off what the facility—which included a garden, a marsh, a miniature ocean, and other habitats—could provide. As a member of the oversight board, JCA was asked to initiate a scientific experiment for the 2-year project inside Biosphere-2. This request came at exactly the same time as when Kim Scribner was conducting his pool and pond experiments with mosquitofish in South Carolina (see the previous abstract). So, Kim and JCA decided likewise to introduce Gambusia *into Biosphere-2 and monitor their cytonuclear dynamics across 2 years. This paper presents the results of that rather peculiar genetic experiment.*

ABSTRACT

Two species of mosquitofish (family Poeciliidae) known to hybridize in nature were introduced into freshwater habitats inside Biosphere-2, and their population genetics were monitored after 2 years. Within four to six generations, nuclear and cytoplasmic markers characteristic of *Gambusia holbrooki* had risen greatly in frequency, although some *Gambusia affinis* alleles and haplotypes were retained primarily in recombinant genotypes, indicative of introgressive hybridization. The temporal cytonuclear dynamics proved similar to population genetic changes observed in replicated experimental hybrid populations outside of Biosphere-2, thus indicating strong directional selection favoring *G. holbrooki* genotypes across a range of environments. When interpreted in the context of the species-specific population demographies observed previously, results suggest that extremely rapid evolution in these zones of secondary contact is attributable primarily to species differences in life history traits.

ADDENDUM

Biosphere-1 is the Earth. Biosphere-2 still exists, although its ownership has shifted over the years. The facility currently is operated by the University of Arizona.

Empirical evaluation of cytonuclear models incorporating genetic drift and tests for neutrality of mtDNA variants: data from experimental *Gambusia* hybrid zones

Scribner, K.T., S. Datta, J. Arnold, and J.C. Avise. 1999. Empirical evaluation of cytonuclear models incorporating genetic drift and tests for neutrality of mtDNA variants: data from experimental Gambusia *hybrid zones. Genetica 105:101–108.*

ANECDOTE OR BACKDROP

Population geneticists have long been preoccupied with the question of whether molecular markers are mostly neutral or, alternatively, whether their evolutionary dynamics are governed by some form(s) of selection. JCA has mostly sidestepped this issue throughout his career, but occasionally its ramifications have been unavoidable. This abstract summarizes one instance in which an inference of nonrandom selection for consistent population genetic trends in the data seems nearly insuperable.

ABSTRACT

Statistical tests of genetic drift and of the neutrality of mtDNA are presented using empirical time series data on multigenerational changes in cytonuclear disequilibria within replicated experimental hybrid populations of two species of live-bearing poeciliid fishes (*Gambusia holbrooki* and *G. affinis*) which were monitored over a period of 2 years (three generations). Cytonuclear disequilibria D1 and D2 (which measure departures from random associations of cytoplasmic and nuclear genotypes) over the three generations of the experiment were non-zero for all replicate populations. For each of five nuclear loci, the observed measures of D1 and D2 were highly concordant between replicates during each generation. Significant departures from expectations were observed after one and two generations. A statistical measure of goodness of fit of observed changes in cytonuclear disequilibria (and implicitly of the neutrality of the mtDNA markers) was calculated for each nuclear locus. When the results for the replicates were combined into an overall test of neutrality, the fit to the random union of zygotes (RUZ) model was rejected for four of the five nuclear loci ($P < 0.05$). A simple genetic drift model does not explain the temporal changes in composite cytonuclear genotypic frequencies. Frequencies of parental *G. holbrooki* mitochondrial alleles and nuclear genotypes exceeded expected values during most time periods, implying some selective advantage for offspring produced by *G. holbrooki* females. An expansion of cytonuclear models to explicitly address questions of genetic drift and neutrality has general relevance to studies of natural populations.

Microsatellite assessment of multiple paternity in natural populations of a live-bearing fish, *Gambusia holbrooki*

Zane, L., W.S. Nelson, A.G. Jones, and J.C. Avise. 1999. Microsatellite assessment of multiple paternity in natural populations of a live-bearing fish, Gambusia holbrooki. **Journal of Evolutionary Biology 12:61–69.**

ANECDOTE OR BACKDROP

One of the earliest molecular studies of genetic paternity in any female-pregnant fish species involved protein electrophoretic analyses that demonstrated moderate rates of multiple paternity in broods of Gambusia. *Here we extended such allozyme analyses by monitoring microsatellite loci in mosquitofish broods, the rationale being that such highly polymorphic markers might yield even higher incidences of multiple paternity. Indeed they did more than 90% of the assayed mosquitofish broods proved to have had more than one sire.*

ABSTRACT

Three polymorphic microsatellite loci were isolated and employed to examine paternity patterns in two natural populations of live-bearing mosquitofish, *Gambusia holbrooki*. Each locus displayed four to five alleles per population in samples of presumably unrelated adults. Nearly 900 embryos from a total of 50 pregnant females were assayed individually, and paternal alleles in each embryo were identified. Counts of paternal alleles, Mendelian segregation patterns, multilocus allelic associations, and genetic relatedness coefficients were employed to estimate the minimum and effective numbers of fathers per brood. At least 90% of the assayed broods were shown to have been fathered by multiple males, a figure substantially higher than previous estimates based on less polymorphic genetic loci. However, the genetic data yield a face-value estimate of only about 2.2 fathers per brood, a number that seems perhaps surprisingly low based on frequencies of attempted copulations by males. Both biological and sampling factors that might bias mean sire counts downward are considered. Although higher sire counts per brood might be obtained from loci with even greater numbers of alleles, little statistical room remains for higher frequency estimates of multiple paternity in *Gambusia*.

Pronounced reproductive skew in a natural population of green swordtails, *Xiphophorus helleri*

Tatarenkov, A., C.I.M. Healey, G.F. Grether, and J.C. Avise. 2008. Pronounced reproductive skew in a natural population of green swordtails, Xiphophorus helleri. *Molecular Ecology 17:4522–4534.*

ANECDOTE OR BACKDROP

This study of genetic parentage in a fish species was another special case in which the relevant populations could be nearly exhaustively sampled (from isolated pools along a rocky stream in Belize). Exhaustive collecting permits far more detailed analyses of reproductive success and reproductive skew than do standard schemes that sample only small subsets of a population. JCA was jealous that his postdoc Andrei Tatarenkov was the one who got to travel to Central America to collect these specimens.

ABSTRACT

For many species in nature, a sire's progeny may be distributed among a few or many dams. This poses logistical challenges—typically much greater across males than across females—for assessing means and variances in mating success (number of mates) and reproductive success (number of progeny). Here we overcome these difficulties by exhaustively analyzing a population of green swordtail fish (*Xiphophorus helleri*) for genetic paternity (and maternity) using a suite of highly polymorphic microsatellite loci. Genetic analyses of 1,476 progeny from 69 pregnant females and 158 candidate sires revealed pronounced skews in male reproductive success both within and among broods. These skews were statistically significant, greater than in females, and correlated in males but not in females with mating success. We also compare the standardized variances in swordtail reproductive success to the few such available estimates for other taxa, notably several mammal species with varied mating systems and degrees of sexual dimorphism. The comparison showed that the opportunity for selection on male *X. helleri* is among the highest yet reported in fishes, and it is intermediate compared to estimates available for mammals. This study is one of a few exhaustive genetic assessments of joint-sex parentage in a natural fish population, and results are relevant to the operation of sexual selection in this sexually dimorphic, high-fecundity species.

ADDENDUM

This is the same species that is popular with aquarists and often sold in pet stores.

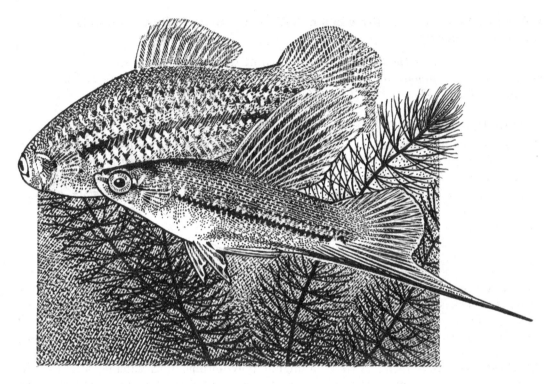

FIGURE 3.4   Green Swordtail, *Xiphophorus helleri.*

Microgeographic population structure of green swordtail fish: genetic differentiation despite abundant migration

Tatarenkov, A., C.I.M. Healey, and J.C. Avise. 2010. Microgeographic population structure of green swordtail fish: genetic differentiation despite abundant migration. **Molecular Ecology 19:257–268.**

ANECDOTE OR BACKDROP

This study was an ancillary follow-up or byproduct of the genetic analyses that had been conducted for the preceding paper on genetic parentage of Green Swordtail fish from Belize.

ABSTRACT

Swordtails (*Xiphophorus*; Poeciliidae) have figured prominently in research on fish mating behaviors, sexual selection, and carcinogenesis, but their population structures and dispersal patterns have been relatively neglected. Using nine microsatellite loci, we estimated genetic differentiation in *Xiphophorus helleri* within and between adjacent streams in Belize. The genetic data were complemented by a tagging study of movement within one stream. In the absence of physical dispersal barriers (waterfalls), population structure followed an isolation by distance (IBD) pattern. Genetic differentiation (F_{ST} up to 0.07) was significant between and within creeks, despite high dispersal in the latter as judged by the tagging data. Such heterogeneity apparently was a result of genetic drift in local demes, due to small population sizes and highly skewed paternity. The IBD pattern was interrupted by waterfalls, boosting F_{ST} above 0.30 between adjacent samples across these barriers. Overall, our results are helpful in understanding the interplay of evolutionary forces and population dynamics in a small fish living in a changeable habitat.

Other Freshwater Fishes

INTRODUCTION

The abstracts in this chapter span the full 40+-year career of JCA. They begin with a comparison of allozyme variation in blind troglobitic forms versus full-eyed epigean populations of a Mexican tetra fish, *Astyanax mexicanus* (this was probably the first multilocus electrophoretic survey of any fish species) and they conclude with a recent examination—based on highly polymorphic microsatellite loci—of genetic parentage in a live-bearing embiotocid fish from northern California. Sandwiched in-between are genetic studies of a wide variety of freshwater fishes with oft-fascinating ecological lifestyles and/or evolutionary peculiarities that make them outstanding fodder for molecular genetic appraisals.

Evolutionary genetics of cave-dwelling fishes of the genus *Astyanax*

Avise, J.C. and R.K. Selander. 1972. Evolutionary genetics of cave-dwelling fishes of the genus Astyanax. *Evolution 26:1–19.*

ANECDOTE OR BACKDROP

This paper was essentially JCA's Masters thesis conducted under the direction of Robert Selander at the University of Texas. A compelling question during that early allozyme era was as follows: Is the magnitude of genetic variation in a population predicted by the magnitude of environmental variation? A positive correlation between ecological heterogeneity and population heterozygosity might suggest that the former actively promotes the latter via spatially varied selection pressures. According to textbooks, caves and the deep sea should be among the most uniform and stable environments on Earth. Our genetic analysis of troglobitic fish was intended to address whether cave-adapted animals show less genetic variation than their close relatives living in surface streams. Although some cave-dwelling populations of Astyanax did indeed prove to show

exceptionally low allozyme heterozygosity, there were good reasons to think that this outcome regis-
tered genetic drift in small populations more so than a uniform environmental selection regime.
This was the first publication of JCA's scientific career, and it got him off to a good start. It was
probably the first multilocus allozyme survey on any fish species, and among the earliest allozyme
studies to document that founder effects and genetic drift can have profound impacts on population
genetic patterns in nature.

ABSTRACT

Allozymic variation in 11 proteins encoded by 17 loci was analyzed in 393 individuals
of the characid fish *Astyanax mexicanus*, representing populations in three cave and six
surface rivers and arroyos in northeastern Mexico. Coefficients of genic similarity among
cave and surface populations averaged 0.82, a value similar to those reported for
conspecific populations of other vertebrates. This degree of similarity presumably
reflects a history of periodic connection between temporarily isolated populations and a
consequent reduction in opportunity for genetic divergence. The genetic information
does not support the assignment of surface and cave populations to different genera or
to different species. Levels of genic variability are high in surface populations, with individuals having 11.2% of their loci in heterozygous state, on the average. In cave populations, variability is absent (Pachon Cave), severely reduced (Los Sabinos Cave, 3.2%), or
less than the average for surface populations (Chica Cave, 7.7%). At 3 of 17 loci, populations in Pachon and Los Sabinos caves are monomorphic or nearly so for different
alleles; and, at a fourth locus, both are monomorphic for an allele not detected in
surface populations. A connection between degree of eye development and allozymic
genotype for the sample from Chica Cave suggests that individuals from the surface are
entering the cave and interbreeding with the eyeless forms. Reduced variability in the
small populations of *Astyanax* in cave pools is attributed primarily to genetic drift.
The eyeless, unpigmented condition is believed to have evolved in whole or part prior
to the present-day subdivision of the populations, which presumably resulted from
recession of floodwaters or shifting drainage patterns. According to a model of periodic
interconnection of caves and intermittent gene flow among cave populations, selection
and other deterministic evolutionary processes could affect most or all of the cave populations as a unit. Owing to the subdivision of aquatic troglobites into small populations,
it is not possible to distinguish the effects of drift and stabilizing or other forms of
selection in reducing genetic variability. An examination of a terrestrial troglobitic beetle
(*Rhadine subterranea*) in two caves near Austin, TX, demonstrated that the presumably
uniform, stable environment of caves does not necessarily impose low degrees of genetic
variability in populations.

ADDENDUM

Blind and unpigmented specimens of Astyanax mexicanus (cave forms of the Mexican Tetra) repre-
sent the same species that is routinely sold in pet stores to freshwater aquarists.

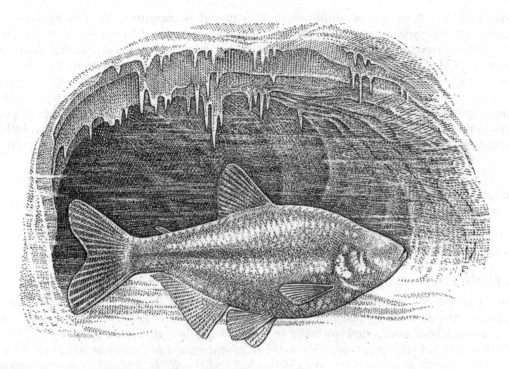

FIGURE 4.1   Blind Cave Tetra, *Astyanax mexicanus*.

Phosphoglucose isomerase gene duplication in the bony fishes: an evolutionary history

Avise, J.C. and G.B. Kitto. 1973. Phosphoglucose isomerase gene duplication in the bony fishes: an evolutionary history. **Biochemical Genetics 8:113–132.**

ANECDOTE OR BACKDROP

During the course of surveying cave populations for genetic variation (see the preceding abstract), JCA stumbled upon molecular evidence for what would prove to be an ancient gene duplication for an important enzyme in the glycolytic pathway of fishes. JCA then studied biochemical properties of these gene products in the laboratory of biochemist Barry Kitto at the University of Texas. JCA also surveyed numerous piscine taxa for the presence versus absence of these duplicated genes. These other fish taxa were collected at the Marine Biological Laboratory in Woods Hole, MA during the summer following JCA's receipt of his Master's degree from the University of Texas.

ABSTRACT

Electrophoretic patterns of phosphoglucose isomerase (PGI) in bony fishes provide strong evidence for a model of genetic control by two independent structural gene loci, most likely resulting from a gene duplication. This model is confirmed by a comparison of kinetic and molecular properties of the PGI homodimers isolated from extracts of the teleost *Astyanax mexicanus*. In addition, in most higher teleosts examined, the PGI enzymes show a regular pattern of tissue distribution, with PGI-2 predominant in muscle, the heterodimer often strongest in the heart, and PGI-1 predominant in liver and other organs. An examination of 53 species of bony fishes belonging to 38 families indicates a widespread occurrence of duplicate PGI loci and an early origin of the gene duplication, perhaps in the Leptolepiformes. The apparent presence of three PGI loci in trout and goldfish exemplifies how new loci can be incorporated into the genome through polyploidization.

Adaptive differentiation with little genic change between two native California minnows

Avise, J.C., J.J. Smith, and F.J. Ayala. 1975. Adaptive differentiation with little genic change between two native California minnows. Evolution 29:411–426.

ANECDOTE OR BACKDROP

This paper was the first part of JCA's Ph.D. dissertation from the University of California at Davis, awarded in 1975. By that time, it was apparent from the broader allozyme literature that even closely related species typically differ in allelic composition at substantial fractions of their protein-coding loci. This paper was noteworthy because it seemed to provide an exception to that rule. Two common minnows native to California had traditionally been placed in separate genera, and yet they proved to be nearly indistinguishable at all surveyed allozyme loci. This study gave one of the first hints that substantial adaptive evolution could take place within the context of relatively little genetic divergence at protein-coding loci.

ABSTRACT

We have studied allelic variation at 24 loci coding for soluble proteins in two presumed species of California minnows—the roach fish *Hesperoleucus symmetricus* and the hitch *Lavinia exilicauda*. Our estimates indicate that they differ at one electrophoretically detectable codon substitution for every 20 loci. This small amount of genetic divergence is unusual in interspecific comparisons. Nonetheless, there is evidence that *Hesperoleucus* and *Lavinia* are different species. They are considerably different in morphological and ecological attributes, and exhibit strong prezygotic isolating barriers. Their distinctness is largely retained also where they are sympatric. The close genic similarity between *Hesperoleucus* and *Lavinia* is due in part to their recent separation from a common ancestor, perhaps during the middle or late Pleistocene. In any case, significant adaptive differentiation leading to speciation has occurred within the context of relatively few genic changes.

ADDENDUM

Notwithstanding their close genic similarity and the fact that they occasionally produce fertile hybrids in nature, taxonomists continue to place the hitch and roach fishes in separate monotypic genera.

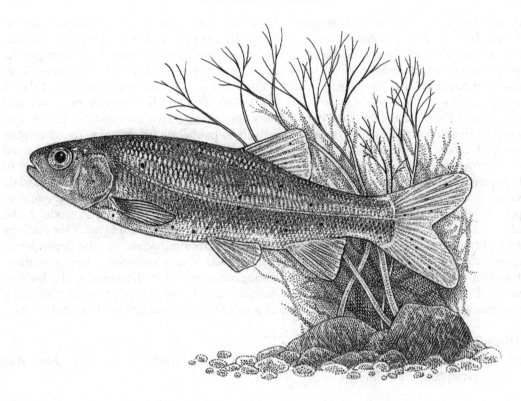

FIGURE 4.2 California Roach Fish, *Hesperoleucus symmetricus.*

Genetic differentiation in speciose versus depauperate phylads: evidence from the California minnows

Avise, J.C. and F.J. Ayala. 1976. Genetic differentiation in speciose versus depauperate phylads: evidence from the California minnows. Evolution *30:46—58.*

ANECDOTE OR BACKDROP

This paper was the second part of JCA's Ph.D. dissertation. It involved genic comparisons between all species and genera of minnows (Cyprinidae) native to California. To collect these specimens, JCA and his fellow students traveled throughout the state, seining or hook-and-line fishing for each species. Unlike popular conceptions of tiny minnows, some of the California cyprinids are large apex predators that can reach a length of 2 feet or more (and can put up a good fight on light tackle).

ABSTRACT

We have examined electrophoretic variation in proteins encoded by 24 gene loci in natural populations of 9 genera of minnows (Cyprinidae) endemic to the freshwaters of California. Average genetic distance, *D*, for all pairwise comparisons among species is 0.57; i.e., about 57 allelic substitutions, on average, are estimated to have occurred for every 100 loci in the separate evolution of any 2 species. At least four genera (*Hesperoleucus, Lavinia, Mylopharodon,* and *Ptychocheilus*) are genetically very similar and have probably evolved from a relatively recent common ancestor. The other genera are less similar; levels of genetic differentiation among them may be fairly representative for the very species-diverse North American minnows. We have also calculated the mean genetic distance among 10 of the 11 known species of the genus *Lepomis*; this is $D = 0.63$. The North American minnows and *Lepomis* are of approximately equal evolutionary age, although the minnows are highly speciose (about 250 species), while *Lepomis* is relatively depauperate. To compare the amount of genetic differentiation in a speciose group versus a depauperate group, we have considered two alternative models: (i) genetic change is a function of time, unrelated to the number of cladogenetic events; (ii) genetic differentiation is proportional to the number of cladogenetic events in the group. According to model 1, the values of D are approximately equal in speciose and depauperate phylads of comparable age. However, according to model 2, the value of D is substantially greater in a speciose than in a depauperate phylad. Our findings of roughly equivalent genetic differentiation in the speciose minnows and the depauperate *Lepomis* support the notion that time since divergence from a common ancestor is more important than the number of intervening cladogenetic events in determining the level of genetic divergence between species. Apparently, the development of reproductive isolating mechanisms *per se* does not involve changes at a substantial proportion of structural genes.

ADDENDUM

Although the taxonomic family to which they belong is huge, in truth there are only a handful of minnow species native to California.

Genetics of plate morphology in an unusual population of threespine sticklebacks (*Gasterosteus aculeatus*)

Avise, J.C. 1976. Genetics of plate morphology in an unusual population of threespine sticklebacks (Gasterosteus aculeatus). **Genetical Research 27:33–46.**

ANECDOTE OR BACKDROP

When JCA began his dissertation work at the University of California at Davis, his original intent was to study genetic patterns in the three-spined stickleback, an interesting species that has morphologically distinct freshwater and marine forms. Much of his first year at U.C. Davis was spent collecting sticklebacks from throughout the state. Unfortunately, most of this effort went for naught because, for unknown reasons, protein-electrophoretic techniques did not work particularly well on the sticklebacks collected. This paper was a last-ditch attempt to salvage at least something from this work, by examining the genetic basis of a morphological trait in captive-reared progeny that JCA had produced by stripping and mixing eggs and sperm from mature adults and then raising the resulting fry in dozens of aquaria.

ABSTRACT

A collection of *Gasterosteus aculeatus* from a single locality (Friant) in Medera County, CA, contains individuals with low and high lateral plate morphology, and very few intermediates. Electrophoretic evidence on protein similarities at 15 genetic loci is compatible with the thesis that members of these two morphs belong to a single inter-breeding population. This thesis is also supported by broods from laboratory crosses between morphs, which segregate for low and high plate counts. Laboratory crosses between Friant fish and those from geographically isolated populations often yield some progeny with intermediate plate counts. The demonstration of significantly different patterns of plate development in intralocality versus interlocality crosses evidences a contrasting genetic basis for plate determination in different populations of sticklebacks.

ADDENDUM

In recent years, several other genetic laboratories have turned freshwater and marine forms of Gasterosteus aculeatus *into an exemplar system for the study of ecological speciation in action.*

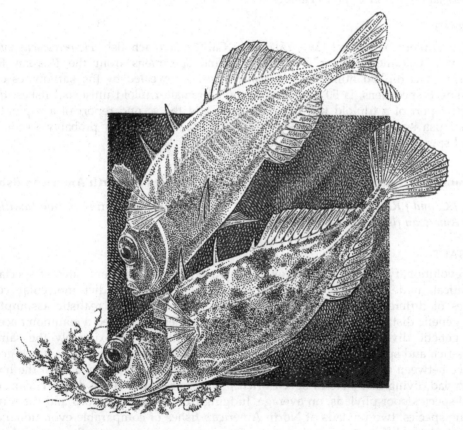

FIGURE 4.3 Three-spined stickleback, *Gasterosteus aculeatus*.

Spontaneous triploidy in the California roach *Hesperoleucus symmetricus* (Pisces: Cyprinidae)

Gold, J.R. and J.C. Avise. 1976. Spontaneous triploidy in the California roach Hesperoleucus symmetricus *(Pisces: Cyprinidae).* Cytogenetics and Cell Genetics 17:144–149.

ANECDOTE OR BACKDROP

While working on his dissertation on California minnows, JCA met and became a close friend with postdoc John R. Gold (JRG) from another U.C. Davis laboratory that specialized on chromosomal analyses of fishes. In that lab, JCA learned how to conduct and interpret chromosomal preparations (karyotypes), after which he and JRG collaborated to produce a series of papers on empirical patterns of chromosomal variation and evolution in a variety of fishes mostly native to California. This abstract and the four that follow it summarize JCA's brief excursion into this realm of chromosomal (as opposed to protein and genic) evolution.

ABSTRACT

A single triploid individual ($3n = 75$) of the California roach fish, *Hesperoleucus symmetricus*, was identified among a sample of nine specimens from the Russian River, California. The diploid number of *H. symmetricus*, as revealed by the karyotypes of the remaining 8 specimens, is 50. Aside from the all-female triploid unisexual fishes, this is the first report of a triploid fish from the wild, and the second report of a triploid in a bisexual fish species. The most likely origin of the triploid was probably fusion of a haploid sperm with an unreduced ovum.

Chromosomal divergence and speciation in two families of North American fishes

Avise, J.C. and J.R. Gold. 1977. Chromosomal divergence and speciation in two families of North American fishes. Evolution 31:1–13.

ABSTRACT

Some evolutionary groups or phylads are characterized by different rates of speciation. Theoretical models have previously been formulated that predict molecular consequences of different speciation rates depending on biologically realistic assumptions. When genetic distance is a function of time since species last shared a common ancestor, mean genetic divergence between living species will be approximately the same in species-rich and species-poor phylads of similar evolutionary age; whereas when genetic distance between species is a function of the number of speciation events in the history of a phylad, living members of species-rich phylads are much more distinct than members of species-poor phylads, on average. Judging from the fossil record and the number of living species, two phylads of North American fishes of comparable evolutionary age are marked by different speciation rates. Speciation events apparently occurred much more frequently in the histories of North American Cyprinidae than in Centrarchidae. In this paper, we continue our search for molecular correlates of speciation rates in these fishes by examining chromosomal cytologies of representative minnows and sunfish. We examined karyotypes of species belonging to nine genera of Cyprinidae and surveyed the literature of published cyprinid karyotypes. Diversity among cyprinid karyotypes is compared to that among representative centrarchids. Karyotypic evolution in both groups is rather conservative. Most North American minnows exhibit $2n = 50$; centromere positions in any species typically show a continuous series from median to terminal. Chromosomes with very short second arms normally comprise less than 20% of the karyotype. Nonetheless, centromeric shifts have repeatedly occurred in the evolutionary histories of these minnows. Most centrarchid species exhibit a karyotype of $2n = 48$, composed entirely of terminal or subterminal chromosomes with very short second arms. Three species have $2n = 46$, in each case presumably the result of a single centric fusion that became fixed in that species. Levels of chromosomal divergence in minnows versus sunfish are discussed in the context of current thought about the relationships between chromosomal evolution, the development of reproductive isolation, and genetic regulation.

Cytogenetic studies in North American minnows (Cyprinidae). I. Karyology of nine California genera

Gold, J.R. and J.C. Avise. 1977. Cytogenetic studies in North American minnows (Cyprinidae). I. Karyology of nine California genera. Copeia 1977:541–549.

ABSTRACT

Karyotypes of nine species representing nine genera of cyprinid fishes inhabiting California were examined. The nine genera, including *Hesperoleucus, Lavinia, Mylopharodon, Pogonichthys, Ptychocheilus, Orthodon, Richardsonius, Gila,* and *Notemigonus,* all have diploid chromosome numbers of 50. *Notemigonus* is the only genus non-native to California having been introduced from the eastern United States. Measurements of centromeric indices suggest differences in fundamental arm number among the genera. In addition, one long chromosome with a distally located centromere was observed in the karyotype of each species and may be of future use in North American cyprinid systematics.

Chromosome cytology in the cutthroat trout series, *Salmo clarki* (Salmonidae)

Gold, J.R., J.C. Avise, and G.A.E. Gall. 1977. Chromosome cytology in the cutthroat trout series, Salmo clarki (Salmonidae). Cytologia 42:377–382.

ABSTRACT

The karyotypes of two subspecies of the cutthroat trout series, *Salmo clarki,* were determined from anterior kidney cells. One subspecies, *S.c. clarki,* the "coastal" cutthroat, has $2n = 68$ chromosomes (36 meta- or submetacentrics and 32 acrocentrics). The other subspecies, *S.c. henshawi,* one of the many forms of "inland" cutthroat, has $2n = 64$ chromosomes (40 meta- or submetacentrics and 24 acrocentrics). The fundamental arm number in both subspecies was estimated at 104. Evidently, chromosomal fusions or dissociations have played a major role in the chromosomal evolution of *S. clarki.*

ADDENDUM

Several subspecies of cutthroat trout are now listed as threatened due to habitat loss in their native ranges and to the introduction of exotic species.

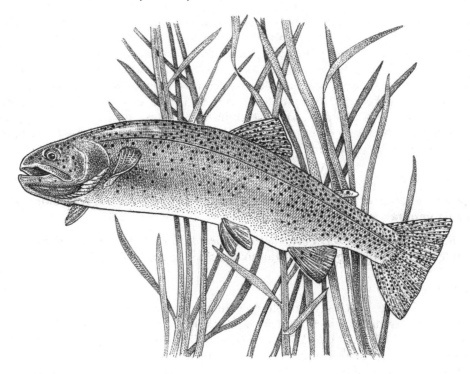

FIGURE 4.4 Cutthroat Trout, *Salmo clarki*.

Cytogenetic studies in North American minnows (Cyprinidae) IV. Somatic polyploidy in *Gila bicolor*

Avise, J.C. and J.R. Gold. 1977. Cytogenetic studies in North American minnows (Cyprinidae) IV. Somatic polyploidy in Gila bicolor. **Canadian Journal of Genetics and Cytology** *19:657–662.*

ABSTRACT

The kidney tissue of a single individual of the California minnow *Gila bicolor* contained polyploid cells in about 1.7% frequency. Chromosome spreads of triploid, tetraploid, hexaploid, octaploid, and dodecaploid cells were observed and may have arisen through endoreduplication of ancestral diploid and triploid cells. The cytological mechanism producing the triploid cells is unknown. The distribution of ploidy in cells of this individual is not random. In particular, cells that have undergone one round of chromosomal increase appear increasingly susceptible to additional rounds of chromosomal gain.

ADDENDUM

The phenomenon of occasional somatic polyploidy via endoreduplication has since been recorded in many species, humans included.

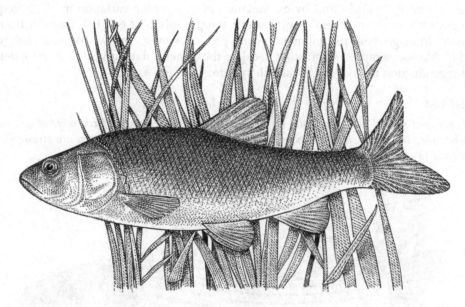

FIGURE 4.5 Tui Chub, *Gila bicolor*.

Genetic analysis of reproduction of hybrid white bass X striped bass in the Savannah River

Avise, J.C. and M.J. Van Den Avyle. 1984. Genetic analysis of reproduction of hybrid white bass X striped bass in the Savannah River. **Transactions of the American Fisheries Society 113:563–570.**

ANECDOTE OR BACKDROP

This is another example of a project sponsored by and conducted in collaboration with Georgia's Department of Natural Resources (DNR), after JCA had obtained a faculty position at the University of Georgia in 1975. It also provides another example (see Chapters 2 and 3*) of how molecular markers can be used to dissect genetic patterns and outcomes in settings of hybridization and possible introgression.*

ABSTRACT

Hatchery-reared F_1 hybrids of male white bass *Morone chrysops* X female striped bass *M. saxatilis* are stocked routinely in the Savannah River drainage by agencies of Georgia and South Carolina. Concern has arisen that these F_1 hybrids themselves may be reproducing, and that possible increases in abundance of F_2 hybrids or backcrosses could disturb white bass and striped bass parental socks through ecological competition or genetic introgression. We identified and employed genetic markers at four protein-electrophoretic systems to search for recombinant genotypes indicative of successful hybrid reproduction. The genetic basic and reliability of the allozyme markers were confirmed by analysis of hatchery-reared F_2 hybrids, and by examination of geographic variation in allele frequencies in populations from several southeastern states. Only 6 of 642 fish assayed from the Savannah drainage had recombinant genotypes. However, in two of four test cases involving *Morone* samples from other locales, the genetic data revealed more extensive hybrid reproduction than in the Savannah system.

ADDENDUM

Hybrids between striped bass and white bass are called "wipers." Produced commercially (by aquaculturists) since the 1980s, they are popular with sport fishermen because they are strong fighters and are good to eat.

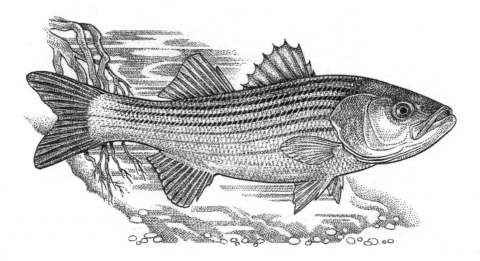

FIGURE 4.6 Striped bass, *Morone saxatilis*.

Molecular zoogeography of freshwater fishes in the southeastern United States

Bermingham, E. and J.C. Avise. 1986. Molecular zoogeography of freshwater fishes in the southeastern United States. **Genetics** *113:939–965.*

ANECDOTE OR BACKDROP

Biff Bermingham was a go-getter graduate student who joined JCA's laboratory at a time when restriction enzymes were just beginning to be used for surveys of geographic variation in vertebrate mitochondrial (mt) DNA sequences. For this project, Biff roamed the southeastern United

States, collecting fish from as many river drainages as possible and by whatever means was necessary. The four species described in this report—three sunfishes and the bowfin—were the only species that consistently showed up in many of these collections. Little did we suspect that any concordant outcomes might emerge from such an eclectic assemblage of piscine taxa, and yet when the molecular data finally were analyzed, some remarkably striking zoogeographic patterns became quite obvious.

ABSTRACT

Restriction fragment length polymorphisms in mitochondrial (mt) DNA were used to reconstruct evolutionary relationships of conspecific populations in four species of freshwater fish—*Amia calva, Lepomis punctatus, L. gulosus,* and *L. microlophus.* A suite of 14–17 endonucleases was employed to assay mtDNAs from 305 specimens collected from 14 river drainages extending from South Carolina to Louisiana. Extensive mtDNA polymorphism was observed within each species. In both phenograms and Wagner parsimony networks, mtDNA clones that were closely related genetically were usually geographically contiguous. Within each species, major mtDNA phylogenetic breaks also distinguished populations from separate geographic regions, demonstrating that dispersal and gene flow have not been sufficient to override geographic influences on population subdivision. Importantly, there were strong patterns of congruence across species in the geographic placements of the mtDNA phylogenetic breaks. Three major boundary regions were characterized by concentrations of phylogenetic discontinuities, and these zones agree well with previously described zoogeographic boundaries identified by a different kind of database—distributional limits of species—suggesting that a common set of historical factors may account for both phenomena. Repeated episodes of eustatic sea level change along a relatively static continental morphology are the likely causes of several patterns of drainage isolation and coalescence, and these are discussed in relation to the genetic data. Overall, results exemplify the positive role that intraspecific genetic analyses may play in historical zoogeographic reconstruction. They also point out the potential inadequacies of any interpretations of population genetic structure that fail to consider the influences of history in shaping that structure.

ADDENDUM

This was probably the first substantive comparative biogeographic survey of mtDNA across multiple codistributed species. Today we would call this a study in comparative phylogeography, but the latter term had not yet been introduced in 1986. When we first began this study on the phylogeography of fishes, we were not even certain that fish would be amenable for mitochondrial assays (most of which previously had been conducted on mammals).

FIGURE 4.7   Bowfin, *Amia calva.*

The molecular basis of a microsatellite null allele from the White Sands pupfish

Jones, A.G., G.A. Stockwell, D. Walker, and J.C. Avise. 1998. The molecular basis of a microsatellite null allele from the White Sands pupfish. Journal of Heredity *89:339–342.*

ANECDOTE OR BACKDROP

As the earlier waves of excitement for allozyme markers and mtDNA markers in population genetics were beginning to subside by the early 1990s, a new wave of interest began to build for another class of polymorphic molecular markers: microsatellite loci. However, as is always true in molecular ecology, each new category of genetic tag should be evaluated for its molecular and evolutionary properties before it is adopted wholesale for particular research tasks. This paper provides an example of one such molecular evaluation of a microsatellite locus.

ABSTRACT

Microsatellite loci were cloned and characterized from the White Sands pupfish (*Cyprinodon tularosa*), a New Mexico state-listed endangered species. One locus exhibited a high-frequency nonamplifying allele localized to a single population. This null allele was PCR amplified by redesign of one of the original primers and multiple individuals homozygous for null as well as for non-null alleles were sequenced using the new primer. These molecular dissections revealed that the original failure to amplify some alleles from this microsatellite locus was due to a 4-bp deletion in one of the original PCR priming sites. Furthermore, the reamplifications revealed five distinct size classes of alleles that had been masquerading as the original null allele. These null alleles did not overlap in length with the non-null alleles, and they also differed consistently by a linked nucleotide substitution. Results suggest that the original null allele (as well as the non-null class) has diversified considerably since its origin and has not recombined frequently with the non-null class of alleles.

ADDENDUM

Null alleles (those that fail to amplify via the polymerase chain reaction) continue to be an unwanted complication in microsatellite surveys of many species. For more on the topic of null alleles, see "Microsatellite null alleles in parentage analysis" in Chapter 16.

Parentage and nest guarding in the tessellated darter (*Etheostoma olmstedi*) assayed by microsatellite markers (Perciformes: Percidae)

DeWoody, J.A., D.E. Fletcher, S.D. Wilkins, and J.C. Avise. 2000. Parentage and nest guarding in the tessellated darter (Etheostoma olmstedi) *assayed by microsatellite markers (Perciformes: Percidae).* **Copeia 2000:740—747.**

ANECDOTE OR BACKDROP

It is not only sunfish (see Chapter 2*) in which males build and tend nests that may contain hundreds or thousands of embryos of uncertain parentage. Many other fish species likewise have such custodial males that may have spawned with multiple females and that occasionally may have been cuckolded by sneaker or other males. To sort out these and other reproductive shenanigans, highly polymorphic microsatellite markers have proved to be extremely powerful for deducing genetic maternity and paternity within each bourgeois fish's nest. When Andrew DeWoody joined JCA's laboratory as a postdoc in the mid-1990s, he was keenly interested in such topics and quickly began to survey several nest-building fish species for genetic parentage, using microsatellite markers that he himself identified (and generated PCR primers for) from scratch.*

ABSTRACT

Parental investment as manifested through extended parental care of young presumably enhances the reproductive success of the custodial parent. In the tessellated darter (*Etheostoma olmstedi*), the primary caregivers are breeding males on the nest. However, prior field observations on nesting darters seem suggestive of behaviors that are more difficult to interpret evolutionarily. These include tending clutches that may have been fertilized by other males, and appropriating nests from smaller courting males. To address such possibilities genetically, we assayed 6 microsatellite loci in 16 nest-tending males and the embryos from their associated clutches. In most cases, a guardian male had sired nearly all of the embryos in his nest. However, in one nest a guardian male had been cuckolded, and in two other nests an attendant male guarded embryos that were not his own, presumably due to nest takeovers. From direct genotypic counts, a mean of at least 3.2 mothers contributed to the progeny in a nest, and computer simulations suggest that the true maternal number may be substantially higher.

ADDENDUM

More than 150 species of Etheostoma darters are native to the streams of eastern North America.

FIGURE 4.8   Tessellated darter, *Etheostoma olmstedi.*

Genetic documentation of filial cannibalism in nature

DeWoody, J.A., D.E. Fletcher, S.D. Wilkins, and J.C. Avise. 2001. Genetic documentation of filial cannibalism in nature. **Proceedings of the National Academy of Sciences USA 98:5090–5092.**

ANECDOTE OR BACKDROP

One day, Andrew DeWoody (a postdoc in JCA's laboratory) thought of a surprising application for genetic parentage analyses. Andrew had been conducting PCR-based microsatellite assays to deduce the genetic parents of tiny embryos in the male-tended nests of several fish species. His standard procedure was to compare the multilocus genotype of each embryo against that of its putative sire (the nest-tender), and make paternity exclusions or inclusions accordingly. Andrew then adjusted his research procedure slightly—by assaying embryos from the stomachs rather than solely from the nests of the custodial males. The result of this peculiar genetic exercise is described in the following abstract and its associated paper.

ABSTRACT

Cannibalism is widespread in natural populations of fishes, where the stomachs of adults frequently contain conspecific juveniles. Furthermore, field observations suggest that guardian males routinely eat offspring from their own nests. However, recent genetic paternity analyses have shown that fish nests often contain embryos not sired by the nest-tending male (because of cuckoldry events, egg thievery, or nest piracy). Such findings, coupled with the fact that several fish species have known capabilities for distinguishing kin from nonkin, raise the possibility that cannibalism by guardian males is directed primarily or exclusively toward unrelated embryos in their nests. Here we test this hypothesis by collecting freshly cannibalized embryos from the stomachs of several nest-tending darter and sunfish males in nature, and determining their genetic parentage using polymorphic microsatellite markers. Our molecular results clearly indicate that guardian males do indeed consume their own genetic offspring, even when unrelated (foster) embryos are present within the nest. These data provide the first genetic documentation of filial cannibalism in nature. Furthermore, they suggest that the phenomenon may result, at least in part, from an inability of guardians to differentiate between kin and nonkin within their own nests.

ADDENDUM

Filial cannibalism apparently occurs in many animals ranging from various insects to some mammals, but the phenomenon seems at face value to be especially common in fish.

Egg mimicry and allopaternal care: two mate-attracting tactics by which nesting striped darter (*Etheostoma virgatum*) males enhance reproductive success

Porter, B.A., A.C. Fiumera, and J.C. Avise. 2002. Egg mimicry and allopaternal care: two mate-attracting tactics by which nesting striped darter (Etheostoma virgatum) *males enhance reproductive success.* **Behavioral Ecology and Sociobiology 51:350–359.**

ANECDOTE OR BACKDROP

Brady Porter and Anthony Fiumera (a postdoc and student, respectively, in JCA's laboratory) were avid connoisseurs of fishes and fish reproductive behaviors in nature. Here they teamed up to genetically disentangle genetic parentage in a freshwater darter species, and thereby contribute to our understanding of the remarkable and widespread phenomenon of egg mimicry in fishes.

ABSTRACT

In a variety of fish species with paternal care of offspring, females prefer to spawn in nests that already contain eggs. This female preference has been hypothesized to explain egg thievery in male sticklebacks, allopaternal care of eggs in minnows, and the evolution of egg-mimicking body features in male cichlids and darters. Here we employ microsatellite-based parentage analyses to evaluate the reproductive success of striped darter (*Etheostoma virgatum*) males that appear to utilize two of these functionally related tactics to entice females to spawn in their nests. In an isolated population (Clear Creek, KY), we observed that breeding males develop conspicuous white spots on their pectoral fins. If these spots are egg mimics, as we suspect, then this represents the fourth independent evolutionary origin of egg mimicry documented to date in darters, the first based on pigmentation (as opposed to physical structures), and the first in which the egg mimics vary greatly in number among males. From direct counts of microsatellite genotypes in clutches of embryos, at least 3.8 females contributed to the progeny within a typical nest, and females tended to spawn preferentially with males that were larger and displayed more egg-mimic spots. In another population (Hurricane Creek, TN) without egg mimics, the multilocus genetic data document that allopaternal care is common, especially among the smallest males who sometimes tend nests containing their own as well as an earlier sire's offspring. Thus, these foster males had adopted egg-containing nests and then successfully spawned with subsequent females. Overall, the genetic data on paternity and maternity, in conjunction with field observations, suggest that egg mimicry and allopaternal care are two mate-attracting reproductive tactics employed by striped darter males to exploit female preferences for spawning in nests with "eggs."

ADDENDUM

For many nest-building fish species, researchers have experimentally documented that females prefer to spawn with males already tending eggs (or at least appearing to tend eggs).

FIGURE 4.9   Striped darter, *Etheostoma virgatum.*

Intensive genetic assessment of the mating system and reproductive success in a semi-closed population of the mottled sculpin, *Cottus bairdi*

Fiumera, A.C., B.A. Porter, G.D. Grossman, and J.C. Avise. 2002. Intensive genetic assessment of the mating system and reproductive success in a semi-closed population of the mottled sculpin, Cottus bairdi. *Molecular Ecology 11:2367–2377.*

ANECDOTE OR BACKDROP

Most genetic surveys of nest-tending fishes sample only a minuscule fraction of the total population, and thus inevitably leave many candidate parents unsampled. In this study and the next, Anthony Fiumera and Brady Porter again joined forces with other associates of the JCA laboratory to genetically assess a special sculpin population that for logistical reasons could be nearly exhaustively collected. The result was a far more complete characterization of genetic parentage in a fish population than is normally possible.

ABSTRACT

Most genetic surveys of parentage in nature sample only a small fraction of the breeding population. Here we apply microsatellite markers to deduce the genetic mating system and assess the reproductive success of females and males in an extensively collected, semi-closed stream population of the mottled sculpin fish, *Cottus bairdi*. In this species, males guard nest rocks where females deposit the eggs for fertilization. The potential exists for both males and females to mate with multiple partners and for males to provide parental care to genetically unrelated offspring. 455 adults and sub-adults, as well as 1,259 offspring from 23 nests, were genotyped at 5 polymorphic microsatellite loci. Multilocus maternal genotypes, deduced via genetic analyses of embryos, were reconstructed for more than 90% of the analyzed nests, thus allowing both male and female reproductive success to be accurately estimated. There was no genetic evidence for cuckoldry, but one nest likely represents a takeover event. Successful males spawned with a mean of 2.8 partners, whereas each female apparently deposited her entire clutch of eggs in a single nest (mean fecundity = 66 eggs/female). On average, genetically deduced sires and dams were captured 1.6 and 9.3 meters from their respective nests, indicating little movement by breeders during the spawning season. Based on a "genetic mark-recapture" estimate, the total number of potentially breeding adults (ca. 570) was an order-of-magnitude larger than genetically based estimates of the effective number of breeders (ca. 54). In addition, significantly fewer eggs *per female* were deposited in single than in multidam nests. Not only were perceived high quality males spawning with multiple partners, but also they were receiving more eggs from each female.

ADDENDUM

The mottled sculpin is a common stream fish throughout much of North America.

FIGURE 4.10 Mottled sculpin, *Cottus bairdi*.

Estimating differential reproductive success from nests of related individuals, with application to a study of the mottled sculpin, *Cottus bairdi*

Jones, B., G.D. Grossman, D.C.I. Walsh, B.A. Porter, J.C. Avise, and A.C. Fiumera. 2007. Estimating differential reproductive success from nests of related individuals, with application to a study of the mottled sculpin, Cottus bairdi. *Genetics 176:2427–2439.*

ABSTRACT

Understanding how variation in reproductive success is related to demography is a critical component in understanding the life history of an organism. Parentage analysis using molecular markers can be used to estimate the reproductive success of different groups of individuals in natural populations. Previous models have been developed for cases where offspring are random samples from the population but these models do not account for the presence of full- and half-sibs commonly found in large clutches of many organisms. Here we develop a model for comparing reproductive success among different groups of individuals that explicitly incorporates within-nest relatedness. Inference for the parameters of the model is done in a Bayesian framework, where we sample from the joint posterior of parental assignments and fertility parameters. We use computer simulations to determine how well our model recovers known parameters and investigate how various data collection scenarios (varying the number of nests or the number of offspring) affects the estimates. We then apply our model to compare reproductive success among different age groups of mottled sculpin, *Cottus bairdi*, from a natural population. We demonstrate that older adults are more likely to contribute to a nest, and that females in the older age groups contribute more eggs to a nest than younger individuals.

Spawning behavior and genetic parentage in the Pirate Perch (*Aphredoderus sayanus*), a fish with an enigmatic reproductive morphology

Fletcher, D.E., E.E. Dakin, B.A. Porter, and J.C. Avise. 2004. Spawning behavior and genetic parentage in the Pirate Perch (Aphredoderus sayanus), *a fish with an enigmatic reproductive morphology.* Copeia 2004:1–10.

ANECDOTE OR BACKDROP

When seining fish from swampy waters in the southeastern United States, people occasionally encounter a small fish that looks extremely peculiar by virtue of the unusual placement of its urogenital pore (under its throat, rather than toward the rear of the fish's body). This morphological configuration sets the Pirate Perch apart from all other fish species in the area, and it also has been a source of much conjecture about its possible adaptive significance. In this study, Dean Fletcher and others from JCA's laboratory use a combination of genetic analyses and field observations to basically solve the longstanding riddle of the Pirate Perch's anterior urogenital opening. The reason d'être proved to be a total surprise.

ABSTRACT

We describe for the first time reproductive behaviors in the Pirate Perch (*Aphredoderus sayanus*), a secretive nocturnal fish whose urogenital opening is positioned far anteriorly, under its throat. Some naturalists had speculated that this peculiar morphological condition might serve to promote egg transfer to the fish's branchial chamber for gill-brooding; others hypothesized that Pirate Perch spawn in the substrate of streams, but offered no adaptive rationale for the odd placement of the fish's urogenital pore. Here we solve the conundrum through a combination of intensive field investigations, underwater filming, and molecular parentage analyses. We show that Pirate Perch spawn in underwater root masses, the first documentation of such nesting behavior in any species of North American fish. Female Pirate Perch thrust their heads and release their eggs into sheltered canals of these masses. Males congregate at these sites and likewise enter the narrow canals headfirst, to release sperm. Thus, the forward-shifted urogenital pore may facilitate spawning under this special nesting circumstance. We found no evidence of extended parental care. Fish formed their own canals or used burrows made by aquatic macro-invertebrates and salamanders. Genetic analyses based on three polymorphic microsatellite loci demonstrate that a total of at least 5—11 sires and dams were the parents of embryos within each of 3 assayed root-mass nests (out of a total of 23 nests found). Males defended the oviposition sites by body-plugging canal entrances after spawning. This and more direct aggressive behaviors by males probably relate to selection pressures imposed by intense competition for fertilization success under these group-spawning conditions.

ADDENDUM

The authors even managed to get some (rather grainy) film footage of the astonishing breeding behavior of the Pirate Perch.

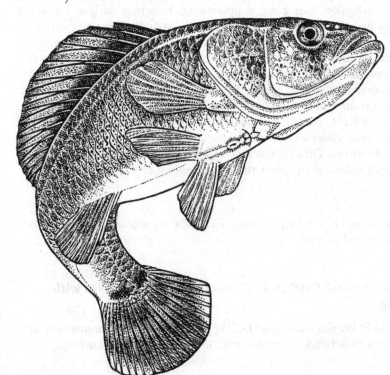

FIGURE 4.11 Pirate Perch, *Aphredoderus sayanus*.

Maximizing offspring production while maintaining genetic diversity in supplemental breeding of highly fecund managed species

Fiumera, A.C., B.A. Porter, G. Looney, M.A. Asmussen, and J.C. Avise. 2004. Maximizing offspring production while maintaining genetic diversity in supplemental breeding of highly fecund managed species. Conservation Biology 18:94–101.

ANECDOTE OR BACKDROP

Most of the research in JCA's laboratory is "pure" (curiosity-driven) as opposed to applied. On occasion, however, an opportunity presents itself to use pure scientific approaches in a more applied context, such as in the conservation of an endangered species. This study provides one such example, involving theoretical analyses of alternative strategies for supplemental breeding of a threatened fish species in Georgia.

ABSTRACT

Supplemental breeding is an intensive population management strategy wherein adults are captured from nature, spawned in controlled settings, and the resulting offspring are later released into the wild. To be effective, supplemental breeding programs require crossing strategies that maximize offspring production while maintaining genetic diversity within each supplemental year class. We used computer simulations to assess the efficacy of different mating designs to jointly maximize offspring production and maintain high levels of genetic diversity (as measured by the effective population size) under a variety of biological conditions particularly relevant to species with high fecundities and external fertilization, such as many fishes. We investigated four basic supplemental breeding designs involving either monogamous pairings or complete factorial designs (where every female is mated to every male and *vice versa*), each with or without the added stipulation that all breeders contribute equally to the total reproductive output. In general, complete factorial designs that did not equalize parental contributions came closest to the goal of maximizing offspring production while still maintaining relatively large effective population sizes. Next, we estimated the effective population size of 10 different supplemental year classes within the breeding program of the Robust Redhorse (*Moxostoma robustum*). Two years failed to produce progeny, whereas successful year classes used partial factorial designs to realize effective sizes ranging from 2 to 26 individuals. On average, a complete factorial design could increase the effective size of each robust redhorse supplemental year class by 19%.

ADDENDUM

The Robust Redhorse was long thought to be extinct, until the 1980s when several specimens were found in rivers of South Carolina and Georgia.

Genetic monogamy in the Channel Catfish, *Ictalurus punctatus*, a species with uniparental nest guarding

Tatarenkov, A., F. Barreto, D.L. Winklelman, and J.C. Avise. 2006. Genetic monogamy in the Channel Catfish, Ictalurus punctatus, *a species with uniparental nest guarding.* Copeia 2006:735–741.

ANECDOTE OR BACKDROP

"Noodling" or handfishing is a peculiar fishing technique in which a fisherman walks or swims in shallow water, searching under structures such as rocks, ledges, and banks for nest-tending catfish. The fish are then grabbed, usually by the lower jaw or opercular opening, pulled from their nest, and placed on a stringer or in a nearby boat. The channel catfishes (plus the egg masses in their nests) that are the subject of this next report were obtained exclusively by noodling in Lake Carl Blackwell, OK. The rest of the study was a bit more conventional—involving microsatellite assessments of biological parentage and the species' genetic mating system.

ABSTRACT

Behavioral observations have suggested that Channel Catfish, *Ictalurus punctatus*, spawn as monogamous pairs and that males alone provide subsequent care to the resulting embryos and fry. However, genetic monogamy is quite uncommon in fish and is not necessarily correctly predicted by apparent social interactions. Here we develop and employ seven microsatellite loci to address biological parentage and the genetic mating system in a natural population of *I. punctatus*. A total of 175 progeny and their respective attendant males were genotyped from 5 nests. Results indicate that each male had mated with only one female in his nest. Additionally, one nest contained a second group of full sibs unrelated to the attendant male and his mate who proved to be the biological parents of all other progeny within that nest. This instance probably represents either a case of nest piracy by the attendant male or perhaps our inadvertent sampling of progeny from two closely adjacent nests. In any event, our findings help confirm a rare suspected example of genetic monogamy in a fish species with uniparental offspring care.

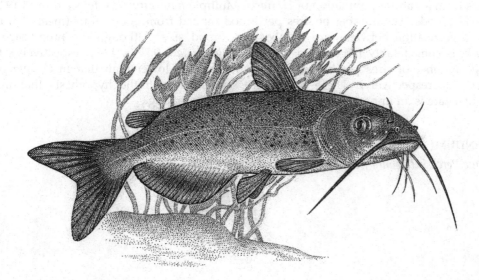

FIGURE 4.12   Channel Catfish, *Ictalurus punctatus*.

Molecular evidence for multiple paternity in a population of the viviparous Tule Perch *Hysterocarpus traksi*

Liu, J., A. Tatarenkov, T.A. O'Rear, P.B. Moyle, and J.C. Avise. 2013. Molecular evidence for multiple paternity in a population of the viviparous Tule Perch Hysterocarpus traksi. *Journal of Heredity 104:217–222.*

ANECDOTE OR BACKDROP

Jason Liu was one of JCA's most recent postdocs. In this study, he again uses microsatellite loci to deduce genetic paternity in a freshwater fish, in this case a female-pregnant cyprinodontid species from northern California. The analytical approach is basically the same as that used to deduce the genetic parentage of embryos in other live-bearing fishes (see Chapters 3 and 5) *and nest-tending fishes (see several earlier abstracts in this chapter).*

ABSTRACT

Population density might be one important variable in determining the degree of multiple paternity. In a previous study, a high level of multiple paternity was detected in the Shiner Perch *Cymatogaster aggregata*, a species with high population density and a high mate encounter rate. The Tule Perch *Hysterocarpus traski* is phylogenetically closely related to *C. aggregata*, but it has relatively lower population density, which may result in distinct patterns of multiple paternity in these two species. To test the hypothesis that mate encounter rate may affect the rate of successful mating, we used polymorphic microsatellite markers to identify multiple paternity in the progeny arrays of 12 pregnant females from a natural population of *H. traski*. Multiple paternity was detected in 11 (92%) of the 12 broods. The number of sires per brood ranged from one to four (mean 2.5) but with no correlation between sire number and brood size. Although the brood size of *H. traski* is considerably larger than that of *C. aggregata* (40.7 vs. 12.9, respectively), the average number of sires per brood in *H. traski* is much lower than that in *C. aggregata* (2.5 vs. 4.6, respectively). These results are consistent with the hypothesis that mate encounter rate is an important factor affecting multiple mating.

ADDENDUM

The Tule Perch is the only freshwater representative of the family Embiotocidae.

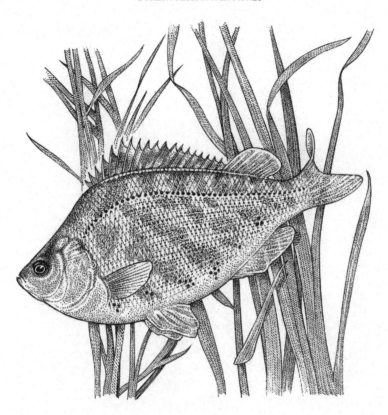

FIGURE 4.13 Tule Perch, *Hysterocarpus traksi.*

Pipefishes and Seahorses (Syngnathidae)

INTRODUCTION

To many women, male pregnancy might seem like a fanciful dream come true (whereas for many men it would be a nightmarish prospect). But for better or worse, the phenomenon does occur (albeit quite rarely) in nature. Indeed, in the 200+ living species of pipefishes and seahorses (family Syngnathidae), males invariably are the pregnant gender, gestating large cohorts of progeny in a ventral brood pouch on their abdomen or tail. This happens when a female transfers her unfertilized eggs, during copulation, to the brood pouch of a male, who fertilizes the eggs internally and then carries and nurtures the resulting embryos for several weeks, until finally giving birth to live young. The whole process is quite analogous to female pregnancy, except that the "conventional" sexual roles of the two genders in effect tend to be reversed.

Male pregnancy has other ramifications, some of which are especially germane to the topics addressed in this chapter. For example, because a pregnant male is physically associated with his biological offspring, polymorphic genetic markers can be recruited to the task of assessing maternity within a clutch (using analytical procedures quite analogous to those used in conventional paternity analyses in humans or other mammals with female pregnancy). Interestingly, however, in the ethological literature male pregnancy alone is not sufficient to categorize a species as being "sex-role reversed." Instead, sex-role reversal is generally characterized as any situation in which the intensity of sexual selection operating on females is greater than the intensity of sexual selection operating on males. By this definition, some male-pregnant species might be sex-role reversed whereas others are not. Genetic parentage analyses can help to clarify this issue also, as several abstracts in the current collection will demonstrate.

Nearly all of the following genetic research on male-pregnant syngnathids was spear-headed by Adam Jones, who produced an impressive array of journal articles during his Ph.D. tenure in JCA's lab during the late 1990s.

Microsatellite analysis of maternity and the mating system in the Gulf pipefish *Syngnathus scovelli*, a species with male pregnancy and sex-role reversal

Jones, A.G. and J.C. Avise. 1997. Microsatellite analysis of maternity and the mating system in the Gulf pipefish Syngnathus scovelli, *a species with male pregnancy and sex-role reversal.* **Molecular Ecology 6:203–213.**

ANECDOTE OR BACKDROP

This was the first of several articles in which Adam Jones summarized his genetic (microsatellite) findings on various pipefish and seahorse species displaying male pregnancy. In this case, the assayed population of Gulf Pipefish proved to be genetically polyandrous, meaning that females often mated successfully with multiple males whereas any pregnant male had mated successfully with more than one female only occasionally.

ABSTRACT

Highly variable microsatellite loci were employed to study the mating system of the sexually dimorphic Gulf pipefish *Syngnathus scovelli*. In this species, like others in the family Syngnathidae, "pregnant" males provide all parental care. Gulf pipefish were collected from one locale in the northern Gulf of Mexico, and internally carried broods of 40 pregnant males were analyzed genetically. By comparing multilocus microsatellite fingerprints for the inferred mothers against expected genotypic distributions from the population sample, it was determined that: (i) only one male had received eggs from more than a single female and (ii) on two separate occasions, two different males had received eggs from the same female. Given the high power to detect multiple matings by males, the first finding indicates that only rarely are individual males impregnated by multiple females during the course of a pregnancy. Conversely, given the lower power to detect multiple matings by females due to sampling constraints, the second finding suggests a high frequency of multiple successful matings by females. Thus, this population of Gulf pipefish displays a polyandrous genetic mating system. The relevance of these genetic findings is discussed with regard to the evolution of secondary sex traits in this species and in other syngnathids.

ADDENDUM

This study would prove to be merely the first of many research papers on mating systems in syngnathid fishes by Adam Jones, who has gone on to build a highly successful academic career on the evolutionary and ecological genetics of these fascinating male-pregnant creatures.

FIGURE 5.1   Gulf pipefish, *Syngnathus scovelli.*

Polygynandry in the dusky pipefish *Syngnathus floridae* revealed by microsatellite DNA markers

Jones, A.G. and J.C. Avise. 1997. Polygynandry in the dusky pipefish Syngnathus floridae *revealed by microsatellite DNA markers.* Evolution 51:1611–1622.

ANECDOTE OR BACKDROP

As for the gulf pipefish (see the previous abstract), pregnant and nonpregnant specimens of the dusky pipefish were collected by dragging a fine-meshed seine through shallow eelgrass pastures along a marine shoreline in northern Florida. As in the previous study, microsatellite loci again were used to deduce genetic parentage in male-carried broods. But unlike in the previous study, in this case both males and females apparently had mated successfully with members of the opposite sex. Thus, the mating system of this species can best be described as polygynandrous, a finding that has ramifications for sexual selection theory.

ABSTRACT

In the dusky pipefish *Syngnathus floridae*, like other species in the family Syngnathidae, "pregnant" males provide all postzygotic care. Male pregnancy has interesting implications for sexual selection theory and the evolution of mating systems. Here, we employ microsatellite markers to describe the genetic mating system of *S. floridae*, compare the outcome to a previous report of genetic polyandry for the Gulf pipefish *S. scovelli*, and consider possible associations between the mating system and degree of sexual dimorphism in these species. Twenty-two pregnant male dusky pipefish from one locale in the northern Gulf of Mexico were analyzed genetically, together with subsamples of 42 embryos from each male's brood pouch. Adult females also were assayed. The genotypes observed in these samples document that: (i) cuckoldry by males did not occur; (ii) males often receive eggs from multiple females during the course of a pregnancy (6 males had 1 mate each, 13 had 2 mates, and 3 had 3 mates); (iii) embryos from different females are segregated spatially within a male's brood pouch; and (iv) a female's clutch of eggs often is divided among more than one male. Thus, the genetic mating system of the dusky pipefish is best described as polygynandrous. The genetic results for *S. floridae* and *S. scovelli* are consistent with a simple model of sexual selection, which predicts that for sex-role-reversed organisms, species with greater degrees of sexual dimorphism are more highly polyandrous.

Microsatellite evidence for monogamy and sex-biased recombination in the Western Australian seahorse *Hippocampus angustus*

Jones, A.G., C. Kvarnemo, G.I. Moore, L.W. Simmons, and J.C. Avise. 1998. Microsatellite evidence for monogamy and sex-biased recombination in the Western Australian seahorse Hippocampus angustus. *Molecular Ecology 7:1497–1505.*

ANECDOTE OR BACKDROP

This was the first of Adam Jones' genetic appraisals of a seahorse species (as opposed to a species of pipefish). Unlike many pipefishes, seahorses generally tend to be monomorphic in morphological features that might otherwise be subject to sexual selection. Consistent with this phenotypic observation, this seahorse population proved to be genetically monogamous.

ABSTRACT

Four polymorphic microsatellite loci were used to assess biological parentage of 453 offspring from 15 pregnant males from a natural population of the Australian seahorse *Hippocampus angustus*. Microsatellite genotypes in the progeny arrays were consistent with a monogamous mating system in which both females and males had a single mate during a male brooding period. Multilocus genotypes implicated four females in the adult population sample as contributors of eggs to the broods of collected males, but there was no evidence for multiple mating by females. Based on genotypic data from the progeny arrays, two loci were linked tightly and the recombination rate appeared to be approximately 10-fold higher in females than in males. The utility of linked loci for parentage analyses is discussed.

ADDENDUM

Globally, more than 50 seahorse fishes reside in the genus Hippocampus. The generic name comes from the ancient Greek "hippos" for horse and "kampos" for sea monster.

FIGURE 5.2   Western Australian seahorse, *Hippocampus angustus*.

The genetic mating system of a sex-role-reversed pipefish (*Syngnathus typhle*): a molecular inquiry

*Jones, A.G., G. Rosenqvist, A. Berglund, and J.C. Avise. 1999. The genetic mating system of a sex-role-reversed pipefish (*Syngnathus typhle): a molecular inquiry.* **Behavioral Ecology and Sociobiology** *46:357–365.*

ANECDOTE OR BACKDROP

This study and the four that follow it describe genetic phenomena in another pipefish species in which the genetic data pointed toward a polygynandrous mating system. They represented Adam Jones' ongoing quest to see whether the genetic mating systems of syngnathids in nature and the laboratory agree with the direction and intensity of sexual selection and sexual dimorphism, as theory predicts they should.

ABSTRACT

In the pipefish *Syngnathus typhle* as in other species of Syngnathidae, developing embryos are reared on the male's ventral surface. Although much laboratory research has been directed toward understanding sexual selection in this sex-role-reversed species, few studies have addressed the mating behavior of *S. typhle* in the wild, and none has capitalized upon the power of molecular genetic assays. Here we present the first direct assessment of the genetic mating system of *S. typhle* in nature. Novel microsatellite loci were cloned and characterized from this species, and employed to assay entire broods from 30 pregnant, field-captured males. Genetic analysis of 1,340 embryos revealed that 1–6 females (mean = 3.1) contributed to each brooded clutch, the highest rate of multiple maternity yet documented in any pipefish. Evidence of multiple mating by females also was detected. Thus, this population of *S. typhle* displays a polygynandrous mating system, a finding consistent with previous field and laboratory observations. Our results, considered in comparison with similar studies of other syngnathid species, provide preliminary support for the hypothesis that the genetic mating system is related to the evolution of sexual dimorphism in the fish family Syngnathidae.

Mate quality influences multiple maternity in the sex-role-reversed pipefish *Syngnathus typhle*

Jones, A.G., G. Rosenqvist, A. Berglund, and J.C. Avise. 2000. Mate quality influences multiple maternity in the sex-role-reversed pipefish Syngnathus typhle. Oikos *90:321–326.*

ABSTRACT

In the pipefish *Syngnathus typhle*, pregnant males provide all parental care, and, consequently, females compete more intensely for mates than do males, a phenomenon defined as sex-role reversal. As the genetic mating system influences the operation of sexual selection, we investigate variation in one phenotypic component of mate quality, female body size, as a possible proximate influence on mating system variation in *S. typhle*. Breeding trials were employed, each consisting of a single receptive male with four adult females. In 10 replicates each, each focal male was paired either with sets of small or with sets of large females. Males were allowed to mate freely, and after several weeks of brood development, maternity of the progeny was resolved using three microsatellite loci. Males with access to small and to large females successfully mated with a mean of 2.1 and 1.3 females respectively, a significant difference. Results indicate that variation in female size can affect the mating system and thereby influence sexual selection in pipefish.

The Bateman gradient and the cause of sexual selection in a sex-role-reversed pipefish

Jones, A.G., G. Rosenqvist, A. Berglund, S.J. Arnold, and J.C. Avise. 2000. The Bateman gradient and the cause of sexual selection in a sex-role-reversed pipefish. Proceedings of the Royal Society of London B *267:677–680.*

ANECDOTE OR BACKDROP

In 1948 (the year of JCA's birth), Angus J. Bateman famously argued that a sex-specific relationship between number of mates and reproductive success was foundational for the premise that sexual selection normally operates more strongly on males than on females. Bateman bolstered his viewpoint with experimental data from Drosophila, showing that males with more mates had much higher reproductive outputs than males with fewer mates, but that the same did not hold for females because they were fecundity-limited regardless of how many mates they had acquired. In the regressions describing these relationships between mate numbers and offspring production, steeper slopes existed for male fruit flies than for female fruit flies. In other words, for sexual species with "standard" sex roles, steeper "Bateman gradients" exist for males than for females. But what about the situation in species (such as male pregnant syngnathids, perhaps) that might be sex-role-reversed? Would Bateman gradients for the two genders be reversed relative to the standard situation? This article by Adam Jones and colleagues addresses this issue. The authors essentially replicate Bateman's experiments, this time using pipefishes in aquaria rather than fruit flies in population cages.

ABSTRACT

As a conspicuous evolutionary mechanism, sexual selection has received much attention from theorists and empiricists. Although the importance of the mating system to sexual selection long has been appreciated, the precise relationship remains obscure. In a classic experimental study based on parentage assessment using visible genetic markers, A. J. Bateman proposed more than 50 years ago that the cause of sexual selection in *Drosophila* is "the stronger correlation, in males (relative to females), between number of mates and fertility (number of progeny)." Half a century later, molecular genetic techniques for assigning parentage now permit mirror-image experimental tests of the "Bateman gradient" using sex-role-reversed species. Here we show that, in the male pregnant pipefish *Syngnathus typhle*, females exhibit a stronger positive association between number of mates and fertility than do males, and that this relationship responds in the predicted fashion to changes in adult sex ratio. These findings give empirical support to Bateman's contention that the relationship between mating success and number of progeny is a primary feature of the genetic mating system affecting the strength and direction of sexual selection.

ADDENDUM

Recently, some authors have raised questions about the details of how Bateman analyzed his Drosophila data. Notwithstanding these concerns, Bateman's general insights are now widely accepted.

Clustered microsatellite mutations in the pipefish *Syngnathus typhle*

Jones, A.G., G. Rosenqvist, A. Berglund, and J.C. Avise. 1999. Clustered microsatellite mutations in the pipefish Syngnathus typhle. *Genetics 152:1057–1063.*

ANECDOTE OR BACKDROP

By definition, clustered de novo *mutations enter a population in groups of identical copies, rather than as singletons. If this sounds like a strange or perhaps even an impossible genetic spectacle, then you need to read the paper from which the following abstract was taken. Apparently, clustered mutations are a very real evolutionary phenomenon, although their biological significance remains open for debate.*

ABSTRACT

Clustered mutations are copies of a mutant allele that enter a population's gene pool together due to replication from a premeiotic germline mutation and distribution to multiple success-ful gametes of an individual. Although the phenomenon has been studied in *Drosophila* and noted in a few other species, the topic has received scant attention despite claims of being of major importance to population genetic theory. Here we capitalize upon the reproductive biology of male pregnant pipefishes to document the occurrence of clustered microsatellite mutations and to estimate their rates and patterns from family data. Among a total of 3,195 embryos genetically screened from 110 families, 40% of the 35 detected *de novo* mutant alleles resided in documented mutational clusters. Most of the microsatellite mutations involved small integer changes in repeat copy number, and they arose in approximately equal frequency in paternal and maternal germlines. These findings extend observations on clus-tered mutations to another organismal group and motivate a broader critique of the mutation cluster phenomenon. They also carry implications for the evolution of microsatellites with respect to mutational models and homoplasy among alleles.

ADDENDUM

In our opinion, clustered mutations continue to be an underappreciated and relatively poorly explored genetic phenomenon.

The measurement of sexual selection using Bateman's principles: an experimental test in the sex-role-reversed pipefish *Syngnathus typhle*

Jones, A.G., G. Rosenqvist, A. Berglund, and J.C. Avise. 2005. The measurement of sexual selection using Bateman's principles: an experimental test in the sex-role-reversed pipefish Syngnathus typhle. **Integrative and Comparative Biology 45:874–884.**

ABSTRACT

Angus J. Bateman's classic 1948 study of sexual selection in *Drosophila melanogaster* exerted a major influence on the development of sexual selection theory by stimulating debate on sex roles, sexual conflict, and related topics. However, considerable disagreement still exists regarding the extent to which "Bateman's principles" are helpful in the study of sexual selection. Here, using the sex-role-reversed pipefish (*Syngnathus typhle*) as a model experimental system in which to address measurements of sexual selection, we test the idea that Bateman's principles provide a useful framework for quantifying and comparing animal mating systems. We bred artificial assemblages of pipefish in the laboratory and used microsatellite markers to resolve genetic parentage under three different sex ratio treatments (female-biased, even, and male-biased) designed to manipulate the expected intensity of sexual selection. Measures of the mating system based on Bateman's principles were calculated and compared to expected changes in the intensity of sexual selection. We also compare results to those from a similar study of Bateman's principles in the rough-skinned newt (*Taricha granulosa*), a species with conventional sex roles. We conclude that measures of the mating system based on Bateman's principles accurately capture the relative intensities of sexual selection in these different treatments and species. By extension, we also therefore conclude that Bateman's principles should generally serve to facilitate comparative studies of sexual selection and mating system evolution in nature.

Monogamous pair bonds and mate switching in the Western Australian seahorse *Hippocampus subelongatus*

Kvarnemo, C., G.I. Moore, A.G. Jones, W.S. Nelson, and J.C. Avise. 2000. Monogamous pair bonds and mate switching in the Western Australian seahorse Hippocampus subelongatus. *Journal of Evolutionary Biology 13:882—888.*

ANECDOTE OR BACKDROP

This was the second of Adam Jones' genetic analyses of seahorses. In this case, the added empirical element was that genetic parentage was assessed across time, through sequential male pregnancies. At the risk of anthropomorphizing, we might say that the genetic data demonstrated that mostly monogamous seahorses nonetheless sometimes divorce and remarry the same or a new partner.

ABSTRACT

Apparently monogamous animals often prove, upon genetic inspection, to mate polygamously. Seahorse males provide care in a brood pouch. An earlier genetic study of the Western Australian seahorse demonstrated that males mate with only one female for each particular brood. Here we investigate whether males remain monogamous in sequential pregnancies during a breeding season. In a natural population, we tagged males and sampled young from two successive broods of 14 males. Microsatellite analyses of parentage revealed that eight males remated with the same female and six with a new female. Thus, in this first study to document long-term genetic monogamy in a seahorse, we show that switches of mates still occur. Polygynous males moved greater distances between broods, and tended to have longer interbrood intervals, than monogamous males, suggesting substantial costs associated with the breaking of pair bonds which may explain the high degree of social monogamy in this fish genus.

The genetic mating system and tests for cuckoldry in a pipefish species in which males fertilize eggs and brood offspring externally

McCoy, E.E., A.G. Jones, and J.C. Avise. 2001. The genetic mating system and tests for cuckoldry in a pipefish species in which males fertilize eggs and brood offspring externally. Molecular Ecology 10:1793—1800.

ANECDOTE OR BACKDROP

By the year 2000, JCA's laboratory had collectively analyzed genetic parentage in thousands of syngnathid embryos taken from pregnant males, yet in no instance had a fry been sired other than by the apparent sire. This makes biological sense because fertilization events in most syngnathids are internal, within the male's enclosed brood pouch. [This absence of stolen fertilizations also stands in sharp contrast to the high levels of cuckoldry often detected in species that fertilize their eggs externally, e.g., in nests (see Chapters 4 and 6).] *However, in a few pipefish species the brood pouch is not fully enclosed, thus potentially opening a window of opportunity for sneaker males to achieve some fertilizations. This next study uses microsatellite markers to address this possibility.*

ABSTRACT

Highly variable microsatellite loci were used to study the mating system of *Nerophis ophidion*, a species of pipefish in which the pregnant males carry embryos on the outside of their body rather than in an enclosed brood pouch. Despite this mode of external fertilization and brooding, otherwise rare in the family Syngnathidae, the genotypes of all embryos proved to be consistent with paternity by the tending male, thus indicating that cuckoldry by sneaker males is rare or nonexistent in this species. *Nerophis ophidion* is a phylogenetic outlier within the Syngnathidae, and its reproductive morphology is thought to be close to the presumed ancestral condition for pipefishes and seahorses. Thus, the current genetic results suggest that the evolutionary elaboration of the enclosed brood pouch elsewhere in the family was probably not in response to selection pressures on pregnant males to avoid fertilization thievery. With regard to maternity assignments, our current genotypic data are consistent with behavioral observations indicating that females sometimes mate with more than one male during a breeding episode, and that each male carries eggs from a single female. Thus, the polyandrous genetic mating system in this species parallels the social mating system, and both are consistent with a more intense sexual selection operating on females, and the elaboration of secondary sexual characters in that gender.

Genetic evidence for extreme polyandry and extraordinary sex-role reversal in a pipefish

Jones, A.G., D. Walker, and J.C. Avise. 2001. Genetic evidence for extreme polyandry and extraordinary sex-role reversal in a pipefish. **Proceedings of the Royal Society of London B 268:2531–2535.**

ANECDOTE OR BACKDROP

This empirical study was another in the long series of works by Adam Jones on the interrelated topics of genetic mating systems, mating behaviors, sexual selection, sexual dimorphism, and Bateman gradients in male pregnant syngnathids.

ABSTRACT

Due to the phenomenon of male pregnancy, the fish family Syngnathidae (seahorses and pipefishes) historically has been considered an archetypal example of a group in which sexual selection should act more strongly on females than on males. However, more recent work has called into question the idea that all species with male pregnancy are sex-role reversed with respect to the intensity of sexual selection. Furthermore, no studies have formally quantified the opportunity for sexual selection in any natural breeding assemblage of pipefishes or seahorses to demonstrate conclusively that sexual selection acts most strongly on females. Here we use a DNA-based study of parentage in the Gulf pipefish, *Syngnathus scovelli*, to show that, indeed, sexual selection acts more strongly on females than on males in this species. Moreover, the Gulf pipefish exhibits classical polyandry with the greatest asymmetry in reproductive roles (as quantified by variances in mating success) between males and females yet documented in any system. Thus, the intensity of sexual selection on females in pipefish rivals that of any other taxon yet studied.

Mating systems and sexual selection in male pregnant pipefishes and seahorses: insights from microsatellite-based studies of maternity

Jones, A.G. and J.C. Avise. 2001. Mating systems and sexual selection in male pregnant pipefishes and seahorses: insights from microsatellite-based studies of maternity. Journal of Heredity 92:150–158.

ANECDOTE OR BACKDROP

This was a review paper summarizing our genetic work on male pregnant species in the family Syngnathidae, published after Adam Jones had graduated with his Ph.D. from the University of Georgia. Adam has since gone on to forge a successful academic career expanding on several genetic horizons opened by his dissertation.

ABSTRACT

In pipefishes and seahorses (family Syngnathidae), the males provide all postzygotic care of offspring by brooding embryos on their ventral surfaces. In some species, this phenomenon of male "pregnancy" results in a reversal of the usual direction of sexual selection, such that females compete more than males for access to mates, and secondary sexual characters evolve in females. Thus, the syngnathids can provide critical tests of theories related to the evolution of sex differences and sexual selection. Microsatellite-based studies of the genetic mating systems of several species of pipefishes and seahorses have provided insights into important aspects of the natural history and evolution of these fishes. First, males of species with completely enclosed pouches have complete confidence of paternity, as might be predicted from parental investment theory for species in which males invest so heavily in offspring. Second, a wide range of genetic mating systems has been documented in nature, including genetic monogamy in a seahorse, polygynandry in two species of pipefish, and polyandry in a third pipefish species. The genetic mating systems appear to be causally related to the intensity of sexual selection, with secondary sex characters evolving most often in females of the more polyandrous species. Third, genetic studies of captive-breeding pipefish suggest that the sexual selection gradient (or Bateman gradient) may be a substantially better method for characterizing the mating system than previously available techniques. Finally, these genetic studies of syngnathid mating systems have led to some general insights into the occurrence of clustered mutations at microsatellite loci, the utility of linked loci in studies of parentage, and the use of parentage data for direct estimation of adult population size.

ADDENDUM

Although several syngnathid species have been genetically assayed in JCA's laboratory, these represent only a small fraction of the more than 200 extant species that reside in this taxonomic family.

Sympatric speciation as a consequence of male pregnancy in seahorses

Jones, A.G., G.I. Moore, C. Kvarnemo, D. Walker, and J.C. Avise. 2003. Sympatric speciation as a consequence of male pregnancy in seahorses. **Proceedings of the National Academy of Sciences USA** *100:6598−6603.*

ANECDOTE OR BACKDROP

Sympatric speciation has been called evolutionary biology's "ugly duckling": always waddling around the field but never quite taking flight or gaining standard scientific acceptance. However, this situation may have changed a bit in recent years, as modes of ecological speciation have attracted renewed interest and several putative examples of sympatric speciation in various taxa have come to light. This paper describes one such possible example from JCA's laboratory. Again, it involves syngnathid seahorses.

ABSTRACT

The phenomenon of male pregnancy in the family Syngnathidae (seahorses, pipefishes, and sea dragons) undeniably has sculpted the course of behavioral evolution in these fishes. Here we explore another potentially important but previously unrecognized consequence of male pregnancy: a predisposition for sympatric speciation. We present new microsatellite data on genetic parentage, which show that seahorses mate size-assortatively in nature. We then develop a quantitative genetic model based on these empirical findings to demonstrate that sympatric speciation can indeed occur under this mating regime in response to weak disruptive selection on body size. We also evaluate phylogenetic evidence bearing on sympatric speciation by asking whether tiny seahorse species often are sister taxa to large sympatric relatives. Overall, our results indicate that sympatric speciation is a possible if not plausible force in the diversification of seahorses, and that assortative mating (in this case the result of male parental care) may warrant broader attention in the speciation process for some other taxonomic groups as well.

ADDENDUM

The possible occurrence and relative incidences of sympatric speciation remain active topics for investigation, in fishes and many other creatures.

6

Other Marine Fishes

INTRODUCTION

It is not only the male-pregnant syngnathids (see Chapter 5) that have fascinating procreative lifestyles; many other marine fishes likewise have peculiar reproductive operations that make them excellent targets for genetic appraisals using molecular markers. For example, in many marine species "bourgeois" males build nests and tend the embryos, and thereby in effect become the "pregnant" gender because they heft a disproportionate share of the parental burden, albeit during an external type of gestation. Such nest guarding also exposes these males to potential cuckoldry by other males who may sneak onto a borgeois male's nest and "steal" some of the fertilization events with the spawning female(s). Such reproductive shenanigans are difficult to document by traditional field observations, but they can become readily apparent when the clutches are subjected to genetic appraisals of biological parentage. By comparing the genotype of each offspring against that of its parental guardian, genetic paternity and maternity can be assessed with relatively little ambiguity, so such genetic appraisals can reveal how many sires and dams contributed their genes to the hundreds or thousands of fry that may be present within a nest. In a few other species of marine fishes, parents likewise attend their young, albeit via mouthbrooding. Some other marine fishes are at the opposite end of the parental care continuum. They sometimes spawn en masse and produce embryos that drift in the oceanic currents for several weeks before settling out to continue their independent life, without ever having seen their parents. Collectively, for all such species with contrasting lifestyles, fascinating questions arise about dispersal, gene flow, and the geographic population structure of a species. For example, one interesting hypothesis speculated that the larvae produced in each single spawning episode might stay together throughout the pelagic phase of the life cycle and settle out together as full-sib social cohorts subject to kin selection. Such questions are well beyond the purview of field observations alone, but they can be addressed and answered using suitable molecular markers. This chapter provides many examples of how molecular techniques have revolutionized our understanding of population genetic processes in the sea.

Mitochondrial DNA differentiation in North Atlantic eels: population genetic consequences of an unusual life history pattern

Avise, J.C., G.S. Helfman, N.C. Saunders, and L.S. Hales. 1986. Mitochondrial DNA differentiation in North Atlantic eels: population genetic consequences of an unusual life history pattern. **Proceedings of the National Academy of Sciences USA** *83:4350–4354.*

ANECDOTE OR BACKDROP

By the mid-1980s, JCA's laboratory had churned out mtDNA zoogeographic surveys on a number of piscine taxa. Each species had proved to harbor extensive mtDNA variation, typically arranged in striking phylogeographical patterns. Now, we were anxious to characterize additional species that might be uniquely favorable for critically testing our emerging hypothesis that historical population demography (which itself is closely linked to each species' historical ecology, behavior, and physical environment) holds the key to understanding contemporary population structure. One such promising species was the American eel, which according to textbooks was a nonpareil example of a broadly distributed panmictic species (due to its catadromous lifestyle that includes extensive migrations between freshwater and marine sites). Given our empirical experience that strong phylogeographic structure is the norm for most vertebrate species, we had expected to find at least moderate genetic subdivision in eels as well, textbook-wisdom notwithstanding. But as this paper details, these eels turned out to be a phylogeographic exception that helped to prove the broader demographic rule.

ABSTRACT

A survey of restriction site polymorphism in the mitochondrial (mt) DNA of the American eel *Anguilla rostrata* showed no genetic divergence among samples from a 4,000-km stretch of North America coastline. Lack of geographic differentiation in mtDNA over such a large area contrasts sharply with results for terrestrial and freshwater vertebrates and is most likely attributable to the extraordinary life history of these catadromous fishes, which involves perhaps a single spawning population in the western tropical mid-Atlantic ocean and subsequent widespread dispersal of larvae by ocean currents. However, samples of the European eel (nominally *Anguilla anguilla*) are highly distinct from *A. rostrata* in mtDNA genotype (distinguishable by 11 of 14 restriction endonucleases), contradicting some previous suggestions that the two forms belong to the same panmictic population. Results of this study emphasize the importance of life history in shaping phylogeographic population structure.

ADDENDUM

More recently, other researchers likewise have surveyed the European eel across its freshwater range, and that catadromous species too has proved to show minimal population genetic structure over vast spatial scales.

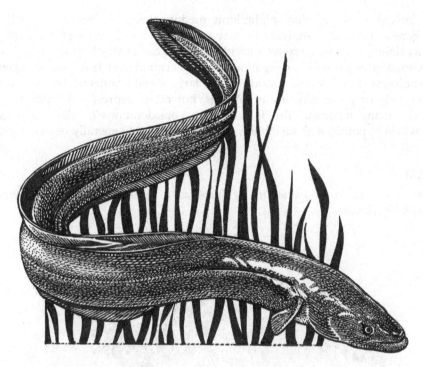

FIGURE 6.1   American eel, *Anguilla rostrata*.

Evaluating kinship of newly settled juveniles within social groups of the coral reef fish, *Anthias squamipinnis*

Avise, J.C. and D.Y. Shapiro. 1986. Evaluating kinship of newly settled juveniles within social groups of the coral reef fish, Anthias squamipinnis. Evolution 40:1051–1059.

ANECDOTE OR BACKDROP

Although this study preceded the microsatellite era by several years, it points out the fact that earlier protein electrophoretic methods also can unveil kinship patterns in nature. Indeed, allozyme data and microsatellite data both entail diploid genotypes at multiple unlinked nuclear loci. One important difference is in the magnitudes of genetic variation. In this paper, allozyme markers nevertheless sufficed to demonstrate that social cohorts in one common reef fish are not composed of close kin. The genetic results probably disappointed my collaborator (Doug Shapiro), a fish behaviorist who wondered whether kin selection might be invoked to explain some of the social behaviors that he had been studying in this species in the Red Sea.

ABSTRACT

It is conventionally assumed that eggs and/or larvae of most coral reef fishes are thoroughly mixed during a pelagic phase, so that juvenile recruits at any particular reef site represent a random sample of the reproductive products entering the local gene pool. However, a recent review of biological factors that might limit mixing raised the testable hypothesis that groups of genetically related individuals may sometimes persist through the pelagic phase and settle as sibling cohorts. Here we provide a critical test of this hypothesis by examining allozyme variation in juvenile aggregations of the serranid reef fish, *Anthias squamipinnis*. Results demonstrate that juvenile cohorts within large social groups in *Anthias* are not composed exclusively or predominantly of siblings, but rather represent a random sample of progeny from many matings. Also included are considerations of allelic and genotypic criteria by which hypotheses about sibling assemblages might generally be evaluated.

ADDENDUM

This common and widespread species in the Indo-Pacific region happens to be a protogynous (female-first) hermaphrodite.

FIGURE 6.2   Lyretail Anthias, *Anthias squamipinnis*.

Geographic population structure and species differences in mitochondrial DNA of mouthbrooding marine catfishes (Ariidae) and demersal spawning toadfishes (Batrachoididae)

Avise, J.C., C.A. Reeb, and N.C. Saunders. 1987. Geographic population structure and species differences in mitochondrial DNA of mouthbrooding marine catfishes (Ariidae) and demersal spawning toadfishes (Batrachoididae). Evolution *41:991−1002.*

ANECDOTE OR BACKDROP

These common species were collected mostly by hook-and-line fishing. Our interest in these two species stemmed from the fact that marine catfishes brood their young orally, whereas toadfishes are demersal (bottom) spawners. The broader goal of this study was to help us develop a broader understanding of how the comparative reproductive ecologies of marine fishes might impact their phylogeographic population structures.

ABSTRACT

Restriction-fragment length polymorphisms in mitochondrial (mt) DNA were used to evaluate population genetic structure and matriarchal phylogeny in four species of marine fishes that lack a pelagic larval stage: the catfishes *Arius felis* and *Bagre marinus*, and the toadfishes *Opsanus tau* and *O. beta*. Thirteen informative restriction enzymes were used to assay mtDNAs from 134 specimens collected from Massachusetts to Louisiana. Considerable genotypic diversity was observed in each species. However, major mtDNA phylogenetic assemblages in catfish and toadfish (as identified in Wagner networks and UPGMA phenograms) exhibited contrasting patterns of geographic distribution: in catfish, distinct mtDNA clades were widespread, while such clades in toadfish tended to be geographically localized. By both the criteria of species' ranges and the geographic pattern of intraspecific mtDNA phylogeny, populations of marine catfish in the western Atlantic have had greater historical interconnectedness than have toadfish. Results are also compared to previously published mtDNA data in freshwater and other marine fishes. Although mtDNA differentiation among conspecific populations of continuously distributed marine fishes is usually lower than that among discontinuously distributed freshwater species inhabiting separate drainages, it is apparent that historical biogeographic factors can importantly influence genetic structure in marine as well as freshwater species.

ADDENDUM

Mouthbrooding and demersal spawning are both quite common phenomena in fishes.

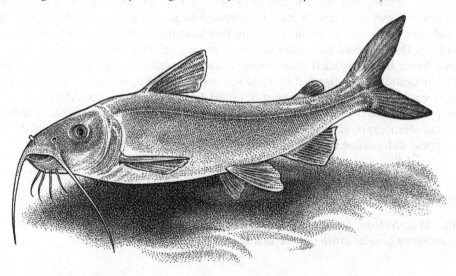

FIGURE 6.3   Hardhead catfish, *Arius felis*.

Current versus historical population sizes in vertebrate species with high gene flow: a comparison based on mitochondrial DNA lineages and inbreeding theory for neutral mutations

Avise, J.C., R.M. Ball. Jr., and J. Arnold. 1988. Current versus historical population sizes in vertebrate species with high gene flow: a comparison based on mitochondrial DNA lineages and inbreeding theory for neutral mutations. Molecular Biology and Evolution 5:331–344.

ANECDOTE OR BACKDROP

Following the seminal allozyme surveys of the mid-1960s, a conceptual irony emerged in the neophyte field of molecular ecology and evolution. For many selectionists, the seminal question became "why is genetic variation so high?" whereas for many neutralists, the seminal question became "why is genetic variation so low?" The latter query stemmed from the fact that polymorphism in many species actually was much lower than predicted by neutrality theory, given the mutation rates and population sizes thought to characterize most species. More than a decade later, analogous questions arose following the seminal empirical surveys of mtDNA. Was the major challenge to explain why mtDNA variation was so high, or so low? This next paper represents one of our early attempts to weigh in on this issue.

ABSTRACT

Using inbreeding theory as applied to neutral alleles inherited maternally, we generate expected probability distributions of times to identity by descent for random pairs of mitochondrial genotypes within a population or within an entire species characterized by high gene flow. For comparisons with these expectations, empirical distributions of times to most recent common ancestry were calculated (by conventional mtDNA clock calibrations) from mtDNA haplotype distances observed within each of three vertebrate species—American eels, hardhead catfish, and red-winged blackbirds. These species were chosen for analysis because census population size in each is currently large and because both genetic and life history data are consistent with the postulate that historical gene flow within these species has been high. The observed molecular distances among mtDNA lineages were 2–3 orders of magnitude lower than predicted from census sizes of breeding females, suggesting that rate of mtDNA evolution is decelerated in these species and/or that long-term effective population size is vastly smaller than present-day population size. Several considerations point to the latter possibility as most likely. The genetic structure of any species is greatly influenced by historical demography; even for species that are currently abundant, mtDNA gene lineages appear to have been channeled through fairly small numbers of ancestors.

The evolutionary genetic status of Icelandic eels

Avise, J.C., W.S. Nelson, J. Arnold, R.K. Koehn, G.C. Williams, and V. Thorsteinsson. 1990. The evolutionary genetic status of Icelandic eels. Evolution 44:1254–1262.

ANECDOTE OR BACKDROP

Hybrid zones normally are situated at the geographical interfaces (or areas of secondary overlap) between pairs of closely related species. This next paper helped to confirm a puzzling exception to this pattern. It involves a hybrid population, in Iceland, between American eels and European eels, both of which were thought to spawn in the Sargasso Sea, thousands of kilometers away. So how in the world do the interspecific hybrids end up in Iceland?

ABSTRACT

The Iceland population of *Anguilla* eels contains an elevated frequency of fish with vertebral numbers lower than those typical of European localities. Several distinct hypotheses have been advanced to account for these morphologically atypical fish: for example, they could represent (i) genetically "pure" American expatriates, (ii) genetically "pure" European types with ontogenetic abnormalities, or (iii) hybrids between American and European forms. Here we critically test these and other possibilities by examining the joint distributions of allozyme markers, mitochondrial DNA markers, and vertebral numbers in Icelandic eels. The particular patterns of association among the genetic and morphological traits demonstrate that the Icelandic population includes, in low frequency, the products of hybridization between American and European eels. Approximately 2–4% of the gene pool in the Iceland eel population is derived from American eel ancestry. This hybrid zone is highly unusual in the biological world, because the mating events in catadromous eels presumably take place thousands of kilometers from where the hybrids are observed as maturing juveniles. The molecular data, in conjunction with geographic distributions, strongly suggest that the differences in migrational behavior and morphology between American and European eels include an important additive genetic component. Evolutionary hypotheses are advanced to account for the original separation of North Atlantic eels into American and European populations, and for the presence of hybrids in Iceland.

ADDENDUM

Subsequent genetic surveys of Icelandic eels (by other laboratories) have further confirmed that hybrids reside on that island and also that the hybrids include generations beyond the F_1.

Genetic structure of Atlantic and Gulf of Mexico populations of sea bass, menhaden, and sturgeon: the influence of zoogeographic factors and life history patterns

Bowen, B.W. and J.C. Avise. 1990. Genetic structure of Atlantic and Gulf of Mexico populations of sea bass, menhaden, and sturgeon: the influence of zoogeographic factors and life history patterns. Marine Biology 107:371–381.

ANECDOTE OR BACKDROP

This study was yet another in our longstanding effort to characterize and interpret phylogeographic patterns in a wide variety of fishes and other vertebrates from the southeastern United States. This particular example involved three marine fish species that have grossly different current-day population sizes, but that also show wide distributions that include both the Atlantic coast and the Gulf of Mexico.

ABSTRACT

To assess the influence of zoogeographic factors and life history parameters (effective population size, generation length, and dispersal) on the evolutionary genetic structure of marine fishes in the southeastern United States, phylogeographic patterns of mitochondrial (mt) DNA were compared between disjunct Atlantic and Gulf of Mexico populations in three coastal marine fishes whose juveniles require an estuarine or freshwater habitat for development. Black sea bass (*Centropristis striata*), menhaden (*Brevoortia tyrannus* and *B. patonus*), and sturgeon (*Acipenser oxyrhynchus*) were collected between 1986 and 1988. All species showed significant haplotype frequency differences between the Atlantic and Gulf, but the magnitude and distribution of mtDNA variation differed greatly among these taxa: sea bass showed little within-region mtDNA polymorphism and a clear phylogenetic distinction between the Atlantic and Gulf; menhaden showed extensive within-region polymorphism and a paraphyletic relationship between Atlantic and Gulf; and sturgeon exhibited very low mtDNA diversity both within regions and overall. Evolutionary effective sizes of the female populations ($N_{f(e)}$) estimated from the mtDNA data ranged from $N_{f(e)} = 50$ (Gulf of Mexico sturgeon) to $N_{f(e)} = 800,000$ (Atlantic menhaden) and showed a strong rank-order agreement with the current-day census sizes of these species. The relationship between $N_{f(e)}$ and the estimated times of divergence among mtDNA lineages (from conventional clock calibrations) predicts the observed phylogenetic distinction between Atlantic and Gulf sea bass, as well as the paraphyletic pattern in menhaden, provided the populations have been separated by the same longstanding zoogeographic barriers thought to have influenced other coastal taxa in the southeastern United States. However, vicariant scenarios alone cannot explain other phylogenetic aspects of the menhaden (and sturgeon) mtDNA data and, for these species, recent gene flow between the Atlantic and Gulf coasts is strongly implicated. The genetic data are relevant to the management and conservation of these species.

ADDENDUM

"Marine" fishes that migrate to spawn in freshwater environments are called anadromous species, in contradistinction to catadromous species that live in freshwater but then migrate to spawn in the sea. Both the sturgeon and menhaden provide examples of anadromous taxa. Both anadromous and catadromous are also sometimes referred to as "diadromous", because they entail two very distinctive phases of a species' lifecycle. For any such taxon, it is problematic as to whether the organism should be referred to as a "marine" or a "freshwater" fish.

FIGURE 6.4   Atlantic sturgeon, *Acipenser oxyrhynchus*.

Pronounced genetic structure of mitochondrial DNA among populations of the circumglobally distributed grey mullet (*Mugil cephalus*)

Crosetti, D., W.S. Nelson, and J.C. Avise. 1994. Pronounced genetic structure of mitochondrial DNA among populations of the circumglobally distributed grey mullet (Mugil cephalus). Journal of Fish Biology 44:47–58.

ANECDOTE OR BACKDROP

Very few fish species are cosmopolitan in their geographic distributions, but one species that comes rather close is the grey mullet, which is found nearly around the world in tropical and subtropical waters. This genetic survey was conducted by a visiting scholar (Donatella Crosetti from Italy) who had a longstanding interest in this commercially important species.

ABSTRACT

Mitochondrial (mt) DNA genotypes of grey mullet (*Mugil cephalus*) from ocean basins around the world were analyzed to estimate the amount of genetic differentiation in this cosmopolitan but mostly coastal-restricted species. Extensive genetic diversity was observed.

Among 115 specimens from 9 locales, 26 different haplotypes were detected using a battery of 13 restriction endonucleases. In phenetic analyses, these haplotypes grouped into seven distinct clusters whose members were in almost perfect accord with the geographic sources of the samples. Thus, contemporary gene flow between the widespread collection locales must be absent or nearly so. Results contrast with the relative uniformity in mtDNA composition previously reported for populations of some circumglobally distributed pelagic fishes, and they demonstrate that some marine fishes with cosmopolitan distributions can exhibit pronounced genetic structure even in the face of morphological conservatism.

ADDENDUM

Mugil cephalus goes by many different common names (such as bully mullet, mangrove mullet, striped mullet, hardgut mullet, flathead mullet) in various countries around the world.

FIGURE 6.5   Grey mullet, *Mugil cephalus.*

A microsatellite assessment of sneaked fertilizations and egg thievery in the fifteenspine stickleback

Jones, A.G., S. Östlund-Nilsson, and J.C. Avise. 1998. A microsatellite assessment of sneaked fertilizations and egg thievery in the fifteenspine stickleback. **Evolution** *52:848–858.*

ANECDOTE OR BACKDROP

In many freshwater fish species with external fertilization, custodial males build and tend nests into which one or more females deposit eggs, a situation that raises many questions about biological maternity and paternity in the large cohorts of embryos within each brood (see Chapters 2 and 4). Similar questions arise for the many marine fish species that likewise construct and monitor their nests. Ethologists have long suspected that several sorts of reproductive shenanigans (alternative reproductive tactics by males or females) might muddy the waters about who truly parented whom. Our microsatellite-based parentage analyses have confirmed the merit of such suspicions by documenting, for example, a relatively frequent occurrence of foster parentage in several nest-tending fish species. This paper provides an example involving marine sticklebacks, for which the genetic data helped to document both fertilization thievery and egg larceny.

ABSTRACT

Attempts by males to steal fertilizations from other males are common in many species. In some sticklebacks, males also are known to steal eggs from the nests of rivals and to carry them back to their own nests. However, the genetic consequences of these nest-raiding behaviors seldom have been investigated. Here we assess genetically the prevalence of sneaked fertilizations and egg stealing, and we describe the mating system in a natural population of the fifteenspine stickleback. Six microsatellite markers were developed and employed to assay a total of 1,307 embryos from 28 nests. Guardian males and all nest-holding males in the local area also were genotyped for 2–6 loci. Analysis of male genotypes and those of embryos revealed that 5 of the 28 nests (18%) contained progeny from sneaked fertilizations, and that 4 of the 24 nests (17%) with resident males contained stolen egg clutches. Comparisons of the composite DNA genotypes of nest-holding males against those of inferred sneakers implicated one nest holder as the sneaker of a nest 7 meters from his own. Also, the genetic data demonstrated that nests of males frequently contain eggs from multiple females. The multilocus genotypes of inferred mothers indicated that females mate with multiple males, sometimes over distances greater than 1 kilometer.

ADDENDUM

The stickleback nest is built above the substrate and is constructed by gluing together plant material using sticky secretions from the kidney. Worldwide, there are more than a dozen stickleback species (family Gasterosteidae), and they are thought to be evolutionary cousins of the pipefishes and seahorses that were described in Chapter 5. Interestingly, sticklebacks lack scales, and instead several of the species have bony armor plates along the sides of the body.

FIGURE 6.6 Fifteen-spined stickleback, *Spinachia spinachia.*

Tests for ancient species flocks based on molecular phylogenetic appraisals of *Sebastes* rockfishes and other marine fishes

Johns, G.C. and J.C. Avise. 1998. Tests for ancient species flocks based on molecular phylogenetic appraisals of Sebastes *rockfishes and other marine fishes.* **Evolution** *52:1135–1146.*

ANECDOTE OR BACKDROP

The famous cichlid species flocks of the African Rift Valley lakes offer quintessential examples of rapid and recent evolutionary radiations of fish species in confined bodies of water. But what can be said about possible adaptive radiations that may have taken place in the more distant past, or perhaps in open bodies of water such as an ocean basin? These are the kinds of questions addressed herein by Glenn Johns, another graduate student in JCA's laboratory. As possible prototype examples, Johns analyzed previously published genetic distances between numerous species of rockfishes and icefishes from their respective environments in the marine realm. It turned out that each of these phylogenetic assemblages could be considered an ancient "species flock."

ABSTRACT

The concept of species flocks has been central to previous interpretations of patterns and processes of explosive species radiations within several groups of freshwater fishes. Here, molecular phylogenies of species-rich *Sebastes* rockfishes from the northeastern Pacific Ocean were used to test predictions of null theoretical models that assume random temporal placements of phylogenetic nodes. Similar appraisals were conducted using molecular data previously published for particular cichlid fishes in Africa that epitomize, by virtue of a rapid and recent radiation of species, the traditional concept of an intralacustrine "species flock." As gauged by the magnitudes of genetic divergence in cytochrome *b* sequences from mitochondrial DNA, as well as in allozymes, most speciation events in the *Sebastes* complex were far more ancient than those in the cichlids. However, statistical tests of nodal placements in the *Sebastes* phylogeny suggest that speciation events in the rockfishes were temporally nonrandom, with significant clustering of cladogenetic events in time. Similar conclusions also apply to an ancient complex of icefishes (within the Notothenioidei) analyzed in the same fashion. Thus, the rockfishes (and icefishes) may be interpreted as ancient species flocks in the marine realm. The analyses exemplified in this report introduce a conceptual and operational approach for extending the concept of species flocks to additional environmental settings and evolutionary timescales.

ADDENDUM

More than 100 extant species of rockfishes reside in the genus Sebastes, and a large percentage of these can be found in the eastern Pacific Ocean.

FIGURE 6.7 Tiger rockfish, *Sebastes nigrocinctus.*

Microsatellite variation in marine, freshwater and anadromous fishes compared with other animals

DeWoody, J.A. and J.C. Avise. 2000. Microsatellite variation in marine, freshwater and anadromous fishes compared with other animals. Journal of Fish Biology 56:461–473.

ANECDOTE OR BACKDROP

This has been a widely cited review paper of comparative levels of multilocus microsatellite varia-tion in fishes that inhabit different environmental regimes and/or display contrasting lifestyles. It stemmed from Andrew DeWoody's penchant for generating and compiling microsatellite informa-tion from a wide variety of fish species.

ABSTRACT

From a total of 524 microsatellite loci considered in nearly 40,000 individuals of 78 species, freshwater fish displayed levels of population genetic variation (mean heterozygosity, $H = 0.46$, and mean numbers of alleles per locus, $a = 7.5$) roughly similar to those of non-piscine animals ($H = 0.58$ and $a = 7.1$). By contrast, local population samples of surveyed marine fish displayed on average significantly higher heterozygosities ($H = 0.79$) and nearly thrice the number of alleles per locus ($a = 20.6$). Anadromous fish proved to be intermediate to marine and freshwater fish in these regards ($H = 0.68$ and $a = 11.3$). Results parallel earlier comparative summaries of allozyme variation in marine, anadromous, and freshwater fishes, and likely are attributable in part to differences in evolutionary effective population sizes typifying species inhabiting these realms.

ADDENDUM

Highly polymorphic microsatellite loci provide powerful molecular markers that are nearly ubiquitously distributed across metazoan animals (not just fishes).

On the number of reproductives contributing to a half-sib progeny array

DeWoody, J.A., Y.D. DeWoody, A.C. Fiumera, and J.C. Avise. 2000. On the number of reproductives contributing to a half-sib progeny array. Genetical Research 75:95–105.

ANECDOTE OR BACKDROP

In addition to gathering empirical microsatellite data on particular fish species, several students in JCA's laboratory have helped to develop statistical and theoretical methodologies for analyzing such data in the context of paternity and maternity assignments. The next two papers provide examples of such analytical approaches.

ABSTRACT

We address various statistical aspects of biological parentage in multi-offspring broods that arose via multiple paternity or multiple maternity and, hence, consist of mixtures of full- and halfsibs. Conditioned on population genetic parameters, computer simulations described herein permit estimation of (i) the mean number of offspring needed to detect all parental gametes in a brood and (ii) the relationship between the number of distinct parental gametes found in a brood and the number of parents. Results are relevant to the design of empirical studies employing molecular markers to assess genetic parentage in polygynous or polyandrous species with large broods, such as are found in many fishes, amphibians, insects, plants, and other groups. The utility of this approach is illustrated using two empirical data sets involving freshwater fishes in which males tend nests with large numbers of embryos.

ADDENDUM

Many statistical approaches (and associated computer programs) are now available for assessing biological parentage from microsatellite or other genetic data. A review can be found in Mol. Ecol. 12:2511−2523 (2003).

Genetic parentage in large half-sib clutches: theoretical estimates and empirical appraisals

DeWoody, J.A., D. Walker, and J.C. Avise. 2000. Genetic parentage in large half-sib clutches: theoretical estimates and empirical appraisals. Genetics 154:1907−1912.

ABSTRACT

Nearly all of the 906 embryos from a male-tended nest of the sand goby (*Pomatoschistus minutus*) were genotyped at two hypervariable microsatellite loci to document conclusively the number of mothers and their relative genetic contributions to the nest. The true number of mothers determined by this nearly exhaustive genetic appraisal was compared to computer simulation treatments based on allele frequencies in the population, assumptions about reproductive skew, and statistical sampling strategies of progeny subsets. The "ground-truthed" appraisal and the theoretical estimates showed good agreement, indicating that for this nest a random sample of about 20 offspring would have sufficed for assessing the true number of biological parents (but not necessarily their relative genetic contributions). Also, a general di-locus matrix procedure is suggested for organizing and interpreting otherwise cumbersome data sets when extremely large numbers of full-sib and half-sib embryos from a nest are genotyped at two or more hypervariable loci.

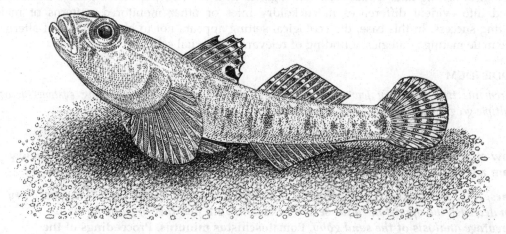

FIGURE 6.8   Sand goby, *Pomatoschistus minutus*.

Surprising similarity of sneaking rates and genetic mating patterns in two populations of sand goby experiencing disparate sexual selection regimes

Jones, A.G., D. Walker, K. Lindström, C. Kvarnemo, and J.C. Avise. 2001. Surprising similarity of sneaking rates and genetic mating patterns in two populations of sand goby experiencing disparate sexual selection regimes. Molecular Ecology 10:461–469.

ANECDOTE OR BACKDROP

The previous paper and the following two all deal with the genetic mating system and sexual selection of the Sand Goby, a small marine fish in which males again guard their nests. In this species, males build the nests under the shells of mussels. During the summer in 1998, Adam Jones from my laboratory visited two of our colleagues in Sweden, where he collected the goby nests by snorkeling in shallow waters and by the use of a hand trawl.

ABSTRACT

Molecular markers have proved extremely useful in resolving mating patterns within individual populations of a number of species, but little is known about how genetic mating systems might vary geographically within a species. Here we use microsatellite markers to compare patterns of sneaked fertilization and mating success in two populations of sand goby (*Pomatoschistus minutus*) that differ dramatically with respect to nest-site density and the documented nature and intensity of sexual selection. At the Tvärminne site in the Baltic Sea, microsatellite genotypes of 17 nest-tending males and mean samples of more than 50 progeny per nest indicated that approximately 35% of the nests contained eggs that had been fertilized by sneaker males. Successful nest holders mated with an average of 3.0 females, and the distribution of mate numbers for these males did not differ significantly from the Poisson expectation. These genetically deduced mating system parameters in the Tvärminne population are remarkably similar to those we had uncovered in sand gobies at a distant site adjoining the North Sea. Thus, pronounced differences in the ecological setting and sexual selection regimes in these two populations have not translated into evident differences in cuckoldry rates or other monitored patterns of male mating success. In this case, the ecological setting appears not to be predictive of alternative male mating strategies, a finding of relevance to sexual selection theory.

ADDENDUM

It remains true that very few molecular studies have assessed genetic mating systems across multiple geographic populations of any species.

How cuckoldry can decrease the opportunity for sexual selection: data and theory from a genetic parentage analysis of the sand goby, *Pomatoschistus minutus*

Jones, A.G., D. Walker, C. Kvarnemo, K. Lindström, and J.C. Avise. 2001. How cuckoldry can decrease the opportunity for sexual selection: data and theory from a genetic parentage analysis of the sand goby, Pomatoschistus minutus. Proceedings of the National Academy of Sciences USA 98:9151–9156.

ABSTRACT

Alternative mating strategies are common in nature and are generally thought to increase the intensity of sexual selection. However, cuckoldry can theoretically decrease the opportunity for sexual selection, particularly in highly polygamous species. We address here the influence of sneaking (fertilization thievery) on the opportunity for sexual selection in the sand goby *Pomatoschistus minutus*, a marine fish species in which males build and defend nests. Our microsatellite-based analysis of the mating system in a natural sand goby population shows high rates of sneaking and multiple mating by males. Sneaker males had fertilized eggs in approximately 50 percent of the assayed nests, and multiple sneakers sometimes fertilized eggs from a single female. Successful males had received eggs from 2 to 6 females per nest (mean = 3.4). We developed a simple mathematical model showing that sneaking in this polygamous sand goby population almost certainly decreases the opportunity for sexual selection, an outcome that contrasts with the usual effects of cuckoldry in socially monogamous animals. These results highlight a more complex and interesting relationship between cuckoldry rates and the intensity of sexual selection than previously assumed in much of the literature on animal mating systems.

ADDENDUM

The basic idea is that cuckoldry can in some circumstances actually decrease the variance in male reproductive success.

Microsatellite variation and differentiation in North Atlantic eels

Mank, J.E. and J.C. Avise. 2003. Microsatellite variation and differentiation in North Atlantic eels. Journal of Heredity *94:310–314.*

ANECDOTE OR BACKDROP

This study, conducted during the microsatellite era, was a logical follow-up to our earlier appraisals of North Atlantic eels by morphological characters and by allozymes and mitochondrial markers (see earlier papers in this chapter). It was our attempt to refine appraisals of geographic population structure in eels, as well as to identify additional markers for analyzing hybridization between the two eel species.

ABSTRACT

We screened 11 populations of American, European, and Icelandic eels (Anguillidae) for allelic variation and genetic divergence at six polymorphic microsatellite loci. Within either of the two recognized *Anguilla* species in the North Atlantic (*rostrata* in the Americas, *anguilla* in Europe), population genetic structure was statistically significant but weak; fully 95% of the total genetic variation was present within geographic locales, rather than distributed among them. The two *Anguilla* species also overlap greatly in allelic frequencies, so the available data proved ineffective for addressing hypotheses about the possible hybrid origins of some Icelandic eels. The overlapping microsatellite profiles contrast with nearly diagnostic species differences documented previously in allozymes and mtDNA. This and similar empirical findings in several other species support theoretical concerns that homoplasy (convergent evolution) in allelic states can compromise the utility of rapidly mutating microsatellite loci for certain types of microevolutionary questions regarding gene flow and species differences.

Cuckoldry rates in the molly miller (*Scartella cristata*; Blennidae), a hole-nesting marine fish with alternative reproductive tactics

*Mackiewicz, M., B.A. Porter, E.A. Dakin, and J.C. Avise. 2005. Cuckoldry rates in the molly miller (*Scartella cristata*; *Blennidae*), a hole-nesting marine fish with alternative reproductive tactics.* **Marine Biology** *148:213–221.*

ANECDOTE OR BACKDROP

This paper addresses genetic parentage in yet another marine species in which males tend nests. In this case, each custodial male is highly territorial in defense of its nest, which is usually the hollow empty shell of a deceased barnacle. Nests and their attendant males were collected as follows: a hollow plastic tube, closed at one end, was placed over an occupied barnacle shell and the custodial male Molly Miller was induced to swim in; the barnacle-nest itself was then scraped off the rock surface by use of a chisel and hammer.

ABSTRACT

Microsatellite markers were developed and employed to assess genetic maternity and paternity of embryos in nest-tended clutches of the Molly Miller (*Scartella cristata*), a marine fish in which alternative reproductive tactics (ARTs) by males were recently described from behavioral and morphological evidence. Genetic data gathered for 1,536 surveyed progeny, from 23 barnacle-nest holes in a single Floridian population, indicate that on average about 5.5 females (range 3–9) contributed to the pool of progeny within a nest. With regard to paternity, the microsatellite data demonstrate that most of the surveyed nests (82.6%) contained at least some embryos that had not been sired by the nest-tending (bourgeois) male, and overall that 12.4% of offspring in the population had been sired via "stolen" fertilizations by other males. These are among the highest values of cuckoldry documented to date in nest-tending fishes, and they support and quantify the notion that the nest-parasitic ART is reproductively quite successful in this species despite what would otherwise seem to be highly defensible nesting sites (the restricted interior space of a barnacle shell). Our estimated cuckoldry rates in this population of the Molly Miller are compared to those previously reported for local populations in other nest-tending fish species, with results discussed in the context of ecological and behavioral variables that may influence relative frequencies of nest parasitism.

ADDENDUM

I have no idea (and can't seem to find) how the Molly Miller Blenny got its common name.

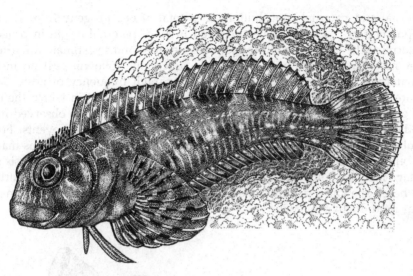

FIGURE 6.9 Molly Miller fish, *Scartella cristata*.

Multiple paternity and extra-group fertilizations in a natural population of California grunion (*Leuresthes tenuis*), a beach-spawning marine fish

Byrne, R.J. and J.C. Avise. 2009. Multiple paternity and extra-group fertilizations in a natural population of California grunion (Leuresthes tenuis), a beach-spawning marine fish. **Marine Biology 156:1681–1690.**

ANECDOTE OR BACKDROP

The grunion is an iconic California marine fish. During the height of the spawning season, at night during a high tide fishermen stalk sandy beaches where the grunion swim ashore en masse to deposit and fertilize their eggs in the sand, amidst waves and swirling masses of fish bodies. We collected grunion the same way that these fishermen do—by hand—except that we (graduate student Rosemary Byrne and JCA) also then shoveled up each nest so that we could conduct genetic parentage analyses on the embryos back in the laboratory. The results presented in this paper and the next are of special interest because grunions in effect are the world's only fish species that spawn "terrestrially."

ABSTRACT

Although individuals in many fish species move to shallow waters to spawn, the California grunion (*Leuresthes tenuis*) is almost unique in its constitutive display of synchronous full-emergence beach spawning. During a spawning event, fish ride large waves onshore to spawn on beach land, where their eggs incubate terrestrially. Here we employ molecular markers to ascertain how this unusual reproductive behavior impacts genetic parentage. We developed and utilized four highly polymorphic microsatellite markers to assess maternal and paternal contributions in a total of 682 progeny from 17 nests of a natural population of *L. tenuis*. Alleles deduced to be of paternal origin in progeny were used to determine the minimum number of sires per nest and to estimate the true number of sires per nest via Bayesian analysis. We document the following: (i) no instances of multiple maternity for progeny within a nest; (ii) a high frequency of nests (88%) with multiple paternity; and (iii) an appreciable fraction of nests (18%) in which the estimated number of genetic sires (as many as 9) proved to be greater than the observed number of male attendants, thus implicating occasional extra-group fertilization events. From these and other observations, we also conclude that spawning behavior in grunions may involve site choice but not explicit mate choice. In addition to providing the first analysis of molecular parentage in a beach-spawning fish, we compare our findings to those reported previously for a beach-spawning arthropod, and we discuss the forces that may be maintaining this peculiar reproductive behavior.

FIGURE 6.10   California grunion, *Leuresthes tenuis*.

Spatiotemporal genetic structure in a protected marine fish, the California grunion (*Leuresthes tenuis*), and relatedness in the genus *Leuresthes*

Byrne R.J., G. Bernardi, and J.C. Avise. 2013. Spatiotemporal genetic structure in a protected marine fish, the California grunion (Leuresthes tenuis), *and relatedness in the genus* Leuresthes. *Journal of Heredity 104:521–531.*

ABSTRACT

The genus *Leuresthes* displays reproductive behavior unique among marine fish in which mature adults synchronously emerge completely out of the water to spawn on beach land. A limited number of sandy beaches that are suitable for these spawning events are present in discontinuous locations along the geographical range of the species, potentially limiting gene flow and the degree of genetic homogeneity between intraspecific populations. Here, we tested for molecular genetic differentiation between three populations of California grunion, *Leuresthes tenuis*, by employing both mitochondrial and nuclear DNA markers. We include temporally diverse sampling to evaluate contemporary as well as temporal divergence and we also analyze one population of Gulf grunion (restricted to the Gulf of California), *Leuresthes sardina*, at the same markers to evaluate the molecular evidence for their separate species distinction. We find no significant differences between temporal samples, but small significant differences among all populations of *L. tenuis*, and unequivocal support for the separate species distinction of *L. sardina*. Genetic data suggest that the Monterey Bay population of *L. tenuis* near the species' most northern range likely represents a relatively recent colonization event from populations along the species' more traditional range south of Point Conception, California. We conclude that both topographical features of the California and Baja California coastlines as well as the grunions' unique reproductive behavior have influenced the genetic structure of the populations.

High degree of multiple paternity in the viviparous shiner perch, *Cymatogaster aggregata*, a fish with long-term female sperm storage

Liu, J.-X. and J.C. Avise. 2011. High degree of multiple paternity in the viviparous shiner perch, Cymatogaster aggregata, *a fish with long-term female sperm storage.* Marine Biology 158:893–901.*

ANECDOTE OR BACKDROP

Although live-bearing in fishes (giving birth to live young following a pregnancy) is notably associated with the Poeciliidae (see Chapter 3) and the Syngnathidae (see Chapter 5), the phenomenon also occurs in several other fish taxa, one example being the marine Shiner Perch (Embiotocidae) that is the subject of this genetic study by postdoc Jason Liu. In this case, pregnant females were collected by hook-and-line-fishing from a rocky jetty in Newport Beach, California, near the University of California at Irvine (where JCA is currently on the faculty).

ABSTRACT

The Shiner Perch (*Cymatogaster aggregata*) exhibits a viviparous reproductive mode and long-term female sperm storage, two biological features that may predispose this fish species for both intense sperm competition and frequent multiple paternity within broods.

To test these hypotheses, we used polymorphic microsatellite markers to identify sires and quantify paternal contributions to the progeny arrays of 27 pregnant females from a natural population of *C. aggregata*. The number of sires per brood ranged from one to eight (mean 4.6), typically with skewed distributions of fertilization success by the fathers but no correlation between sire number and brood size. The extraordinarily high incidences of multiple paternity in this species probably are due in part to high rates of mate encounter, but selection pressures related to the avoidance of maternal-fetal incompatibility may further have promoted the evolution of polyandrous mating behaviors in this female pregnant species. Our genetic data are consistent with the hypothesis that viviparity, long-term sperm storage, and extreme polyandry are interrelated reproductive phenomena that should promote the evolution of postcopulatory sperm competition and/or cryptic female choice in these fishes.

ADDENDUM

The Shiner Perch is one of about 20 small live-bearing embiotocid fishes that mostly inhabit shallow marine waters of the northern Pacific basin.

FIGURE 6.11   Shiner Perch, *Cymatogaster aggregata*.

Effects of Pleistocene climatic fluctuations on the phylogeographic and demographic histories of Pacific Herring (*Clupea pallasi*)

*Liu, J.-X., A. Tatarenkov, T.D. Beacham, V. Gorbachev, S. Wildes, and J.C. Avise. 2011. Effects of Pleistocene climatic fluctuations on the phylogeographic and demographic histories of Pacific Herring (*Clupea pallasi*). Molecular Ecology 20:3879–3893.*

ANECDOTE OR BACKDROP

Herring are a super-abundant and commercially important marine fish in the northern Pacific basin. Here Jason Liu and coauthors used mtDNA sequences from many extant populations to reconstruct the evolutionary and demographic histories of the Pacific Herring in this marine region. The fish were obtained from commercial fishing or research vessels that targeted spawning and prespawning aggregations in nearshore waters.

ABSTRACT

We gathered mitochondrial DNA sequences from the control region and the cytochrome b gene from 935 specimens of Pacific Herring collected from 20 nearshore localities spanning the species extensive range along the North Pacific coastlines of Asia and North America. Using a variety of phylogenetic methods, coalescent reasoning, and molecular dating, all interpreted in conjunction with paleoclimatic and physiographic evidence, we deduce that the genetic makeup of extant populations of *C. pallasi* was shaped to a large extent by Pleistocene environments that impacted the historical demography of this species. For example, a deep genealogical split that cleanly distinguishes populations in the western versus eastern North Pacific probably originated approximately 1 million years ago as a population separation that was associated with a glacial cycle that drove the species southward and vicariantly isolated two ancestral populations in Asia and North America. Another deep genealogical split that dates to the same approximate time frame may have involved a vicariant isolation of a third herring lineage originally in the Gulf of California. In addition to these major phylogeographic separations, coalescent analyses allowed us to deduce the timing and magnitude of several population expansions and contractions within each of the three major evolutionary lineages of *C. pallasi*. Overall, our results dramatize how historical changes in the physical environment have profoundly shaped many genetic features of extant biotas.

ADDENDUM

Notwithstanding its usual abundance, populations of the Pacific Herring can also fluctuate dramatically. For example, much of the North Pacific fishery collapsed in 1993.

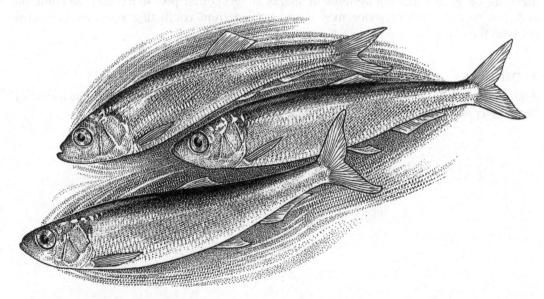

FIGURE 6.12 Pacific Herring, *Clupea pallasi*.

Genetic mating system of the brown smoothhound shark (*Mustelus henlei*), including a literature review of multiple paternity in other elasmobranch species

Byrne, R.J. and J.C. Avise. 2012. Genetic mating system of the brown smoothhound shark (Mustelus henlei), including a literature review of multiple paternity in other elasmobranch species. **Marine Biology 159:749–756.**

ANECDOTE OR BACKDROP

Her genetic papers on grunion fish (see two earlier abstracts in this chapter) were not quite enough to earn Rosemary Byrne her Ph.D. degree, so as an ancillary project she also examined genetic paternity in a viviparous shark species. Tissue samples from pregnant females and their litters of pups were obtained from local commercial fishermen in Baja California, Mexico.

ABSTRACT

Although an understanding of mating systems is thought to be an important component of long-term population management, these life history characteristics are poorly known in sharks. Here we employ polymorphic microsatellite markers to test for the occurrence and prevalence of multiple paternity in a population of the brown smoothhound shark, *Mustelus henlei*. We analyzed litters from 14 females sampled from the Pacific coast of Baja California Sur. The minimum number of sires ranged from 1 to 3 with an average of 2.3 sires per litter. Regression analyses did not indicate a relationship between female body size and number of sires, or female body size and size of the litter. A review of the existing literature on genetic mating systems in sharks suggests that polyandry may be common and that reproductive behavior may have evolved from conflicting selection pressures between the sexes.

ADDENDUM

Approximately 60% of the nearly 1,000 extant species of sharks and rays (cartilaginous fishes) likewise give birth to live young.

Mangrove Rivulus

INTRODUCTION

Most of the research conducted in JCA's lab involves comparisons of genetic or phylogenetic features across a wide diversity of creatures. This typically means that relatively little effort is expended on any one species. But occasionally a single species becomes of sufficient interest to warrant multiple genetic analyses (and publications) over an extended period of time. Four such examples already discussed in this book include the bluegill sunfish, *Lepomis macrochirus* (see Chapter 2), the *Poeciliopsis* complex of unisexual fishes (see Chapter 3), particular pipefish species (see Chapter 5), and North Atlantic eels (see Chapter 6). Another such example involves the remarkable creature described in this chapter—the Mangrove Rivulus, *Kryptolebias marmoratus*. This inch-long species of American killifish (Cyprinodontidae) is special in several regards. Among vertebrate species, it is the world's only known animal that routinely self-fertilizes. Inside each hermaphroditic fish is an "ovotestis" organ that simultaneously produces eggs and sperm. These gametes then unite within the fish's body to produce zygotes that are expelled to the outside world. Also essentially unique among vertebrates is the fact that pure males also exist in *K. marmoratus*, reaching substantial frequencies in some populations. Thus, some populations of this species are actually "androdioecious." Furthermore, the males sometimes mediate outcross events with the hermaphrodites, thereby unleashing extreme recombinational potential among the resulting progeny. These outcross events apparently transpire when a hermaphrodite occasionally sheds a few unfertilized ova that are then fertilized externally by a male's sperm. All of this means that the Mangrove Rivulus actually displays a "mixed-mating" system of occasional outcrossing against a backdrop of predominant selfing. Thus, self-fertilization continued generation after generation leads to a quasi-clonal population genetic structure that occasional outcrossing then tends to dismantle. All of these facts, and many others, have emerged from extensive molecular genetic investigations that have been conducted on this amazing little fish. The abstracts in this chapter summarize some of these studies.

J.C. Avise: Sketches of Nature.
http://dx.doi.org/10.1016/B978-0-12-801945-0.00007-3

Microsatellite documentation of male-mediated outcrossing between inbred laboratory strains of the self-fertilizing mangrove killifish (*Kryptolebias marmoratus*)

Mackiewicz, M., A. Tatarenkov, A. Perry, J.R. Martin, D.F. Elder, Jr., D.L. Bechler, and J.C. Avise. 2006. Microsatellite documentation of male-mediated outcrossing between inbred laboratory strains of the self-fertilizing mangrove killifish (Kryptolebias marmoratus). **Journal of Heredity 97:508–513.**

ANECDOTE OR BACKDROP

This was our first of many papers on the population genetics of the Mangrove Rivulus (Kmar), conducted as a part of the dissertation project of graduate student Mark Mackiewicz (and in collaboration with several other colleagues who specialized on this species). The many microsatellite markers that Mark developed for this in initial study were to prove useful for follow-up studies of Kmar in JCA's lab across the ensuing decade.

ABSTRACT

Primers for 36 microsatellite loci were developed and employed to characterize genetic stocks and detect possible outcrossing between highly inbred laboratory strains of the self-fertilizing mangrove killifish, *Kryptolebias marmoratus*. From attempted crosses involving hermaphrodites from particular geographic strains and gonochoristic males from others, 2 among a total of 32 surveyed progeny (6.2%) displayed multilocus heterozygosity clearly indicative of interstrain gametic syngamy. One of these outcross hybrids was allowed to resume self-fertilization, and microsatellite assays of progeny showed that heterozygosity decreased by approximately 50% after one generation, as expected. Although populations of *K. marmoratus* consist mostly of synchronous hermaphrodites with efficient mechanisms of internal self-fertilization, these laboratory findings experimentally confirm that conspecific males can mediate occasional outcross events and that this process can release extensive genic heterozygosity.

ADDENDUM

The Mangrove Rivulus formerly went by the name Rivulus marmoratus, the Mangrove Killifish. This changing nomenclature may confuse some researchers who may have read in textbooks that Rivulus marmoratus was the world's only self-fertilizing hermaphroditic vertebrate. The Mangrove Rivulus does indeed live in mangrove forests, from southern Florida and throughout the Caribbean region, south to southern Brazil. It can survive for weeks out of water, and can move from pool to pool across the wet forest floor by flipping and flopping.

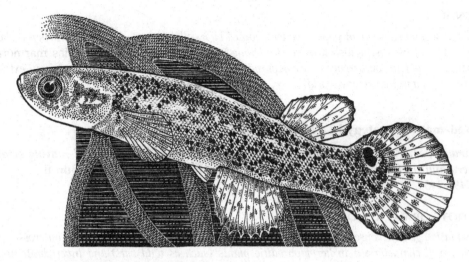

FIGURE 7.1   Mangrove Rivulus, *Kryptolebias marmoratus*.

Extensive outcrossing and androdioecy in a vertebrate species that otherwise reproduces as a self-fertilizing hermaphrodite

Mackiewicz, M., A. Tatarenkov, D.S. Taylor, B.J. Turner, and J.C. Avise. 2006. Extensive outcrossing and androdioecy in a vertebrate species that otherwise reproduces as a self-fertilizing hermaphrodite. **Proceedings of the National Academy of Sciences USA 103:9924–9928.**

ANECDOTE OR BACKDROP

Previous reports from the literature had shown that a population of Kmar from Belize included males in moderate frequency, in addition to the usual array of hermaphroditic specimens. This study demonstrated the dramatic genetic impact that these males have on the population genetics of this species. The fish for this and subsequent studies were collected from Belize by some of our colleagues and coauthors.

ABSTRACT

The mangrove killifish (*Kryptolebias marmoratus*) is the only vertebrate known to be capable of self-fertilization. Its gonad is typically an ovotestis that simultaneously produces eggs and sperm, and fertilization is internal. Although most populations of this species consist primarily or exclusively of hermaphroditic individuals, gonochoristic males occur at about 20% frequency in a natural population at Twin Cays, Belize. Here we use a battery of 36 microsatellite loci to document a striking genetic pattern—high intraspecimen heterozygosities and low within-population linkage disequilibria—that differs qualitatively from the highly homozygous (or "clonal") genetic architecture characteristic of killifish populations previously studied in Florida, where males are much rarer. These findings document that outcrossing (probably between gonochoristic males and hermaphrodites) is common at the Belize site, and more importantly they demonstrate the dramatic impact that functional androdioecy can have on the population genetic architecture of this reproductively unique vertebrate species.

ADDENDUM

In this and several subsequent papers we were joined by a collaborator—Dr. Bruce Turner—who had devoted much of his distinguished career to detailed genetic studies of Kryptolebias marmoratus. *Androdioecy (the joint occurrence of hermaphrodites and pure males within a species) is an extremely rare phenomenon in the organic world.*

A mixed-mating strategy in a hermaphroditic vertebrate

Mackiewicz, M., A. Tatarenkov, B.J. Turner, and J.C. Avise. 2006. A mixed-mating strategy in a hermaphroditic vertebrate. **Proceedings of the Royal Society of London B 273:2449–2452.**

ANECDOTE OR BACKDROP

The population genetic consequences of mixed-mating systems (involving selfing plus outcrossing) had long been well characterized in hermaphroditic plants (such as wild oats) and invertebrate animals (such as some snails), but no such examples were known for vertebrates. This study rectified that shortcoming.

ABSTRACT

Mixed-mating systems, in which hermaphrodites can either self-fertilize or outcross, are common in many species of plants and invertebrates and have been informative models for studying the selective forces that can maintain both inbreeding and outbreeding in populations. Here, we document a remarkable instance of evolutionary convergence to an analogous mixed-mating system in a vertebrate, the mangrove killifish (*Kryptolebias marmoratus*). In this androdioecious species, most individuals are simultaneous hermaphrodites that characteristically self-fertilize resulting in local populations that consist of nearly homozygous lines. Most demes are also genetically diverse, an observation traditionally attributed to *de novo* mutation coupled with high levels of intersite migration. However, data presented here, from a survey of 35 microsatellite loci in Floridian populations, show that genotypic diversity also stems proximally from occasional outcross events that release "explosions" of transient recombinant variation. The result is a local population genetic pattern (of extensive genotypic variety despite low but highly heterogeneous intraindividual heterozygosities) that differs qualitatively from genetic architectures known in any other vertebrate species. Advantages of a mixed-mating strategy in *K. marmoratus* probably relate to this fish's solitary lifestyle and ability to colonize new habitats.

Strong population structure despite evidence of recent migration in a selfing hermaphroditic vertebrate, the mangrove killifish (*Kryptolebias marmoratus*)

*Tatarenkov, A., H. Gao, M. Mackiewicz, D.S. Taylor, B.J. Turner, and J.C. Avise. 2007. Strong population structure despite evidence of recent migration in a selfing hermaphroditic vertebrate, the mangrove killifish (*Kryptolebias marmoratus*).* **Molecular Ecology 16:2701–2711.**

ANECDOTE OR BACKDROP

Russian-born Dr. Andrei Tatarenkov has been a long-term collaborator and manager of JCA's laboratory. In this study and several others that fill out this chapter, he takes the lead in using various classes of molecular markers to further characterize population genetic and evolutionary patterns and processes in the Kmar complex of hermaphroditic fishes.

ABSTRACT

We employ a battery of 33 polymorphic microsatellite loci to describe geographic population structure of the Mangrove Killifish (*Kryptolebias marmoratus*), the only vertebrate species known to have a mixed-mating system of selfing and outcrossing. Significant population genetic structure was detected at spatial scales ranging from tens to hundreds of kilometers in Florida, Belize, and the Bahamas. The wealth of genotypic information, coupled with the inbred nature of most killifish lineages due to predominant selfing, also permitted treatments of individual fish as units of analysis. Genetic clustering algorithms, neighbor-joining trees, factorial correspondence, and related analysis methods all earmarked particular killifish specimens as products of recent outcross events that could often be provisionally linked to specific migration events. Although *de novo* mutation remains the ultimate source of genetic diversity in *K. marmoratus*, we show that interlocality dispersal and outcross-mediated genetic recombination (and probably genetic drift as well) play key proximate roles in generating and shaping local "clonal" variety and "clonal" dynamics in this species.

Rapid concerted evolution in animal mitochondrial DNA

Tatarenkov, A. and J.C. Avise. 2007. Rapid concerted evolution in animal mitochondrial DNA. Proceedings of the Royal Society of London B 274:1795–1798.

ANECDOTE OR BACKDROP

"Concerted" evolution refers to the evolution "in concert" of multiple copies within a family of closely related genetic loci. It implies that such loci somehow mechanistically interact during evolution in such a way that their nucleotide sequences evolve together rather than fully independently. Traditionally, concerted evolution had been studied in nuclear genomes, where recombinational mechanisms are rampant. Thus it was to our great surprise that we uncovered a dramatic empirical case of concerted evolution in animal mtDNA, which conventionally had been perceived to lack all recombinational mechanisms.

ABSTRACT

Recombinational genetic processes are thought to be rare in the uniparentally inherited mitochondrial (mt) DNA molecules of vertebrates and other animals. Here, however, we document extremely rapid concerted microevolution, probably mediated by frequent gene conversion events, of duplicated sequences in the mtDNA control region of mangrove killifishes. In local populations, genetic distances between paralogous loci within an individual were typically smaller (and often zero) than genetic distances between orthologous loci in different specimens. These findings call for the recognition of concerted evolution as a microevolutionary process and gene conversion as a likely recombinational force in animal mtDNA. The previously unsuspected power of these molecular phenomena could greatly impact mtDNA dynamics within germ cell lineages and in local animal populations.

Long-term retention of self-fertilization in a fish clade

Tatarenkov, A., S.M.Q. Lima, D.S. Taylor, and J.C. Avise. 2009. Long-term retention of self-fertilization in a fish clade. **Proceedings of the National Academy of Sciences USA** *106:14456−14459.*

ANECDOTE OR BACKDROP

This study was our first attempt to place the population genetic processes of Kmar into a broader evolutionary or phylogenetic context by similarly analyzing some of Kmar's closely related species.

ABSTRACT

Among vertebrate animals, only the mangrove rivulus (*Kryptolebias marmoratus*) was known to self-fertilize. Here we use microsatellite analyses to document a high selfing rate (97%) in a related nominal species, *K. ocellatus*, which likewise is androdioecious (populations consist of males and hermaphrodites). In contrast, we find no evidence of self-fertilization in *K. caudomarginatus* (an androdioecious species closely related to the *marmoratus-ocellatus* clade) or in *K. brasiliensis* (a dioecious outgroup). These findings indicate that the initiation of self-fertilization predated the origin of the *marmoratus/ocellatus* clade. From mitochondrial DNA sequences and microsatellite data, we document a substantial genetic distance between *K. marmoratus* and *K. ocellatus*, implying that the selfing capacity has persisted in these fishes for at least several hundred thousand years.

Genetic composition of laboratory stocks of the self-fertilizing fish *Kryptolebias marmoratus*: a valuable resource for experimental research

Tatarenkov, A., B.C. Ring, J.F. Elder, D.L. Bechler, and J.C. Avise. 2010. Genetic composition of laboratory stocks of the self-fertilizing fish Kryptolebias marmoratus: *a valuable resource for experimental research.* **PLoS One** *5(9):e12863.*

ANECDOTE OR BACKDROP

The selfing proclivity and clonal population genetic structure of Kmar had long made this species a favored animal for replication-based experimental research in several laboratories around the world. Here we genetically analyzed these laboratory stocks and demonstrate that several of them had been misidentified or mislabeled over the years.

ABSTRACT

The hermaphroditic Mangrove Killifish, *Kryptolebias marmoratus*, is the world's only vertebrate that routinely self-fertilizes. As such, highly inbred and presumably isogenic "clonal" lineages of this androdioecious species have long been maintained in several laboratories and used in a wide variety of experiments that require genetically uniform vertebrate specimens. Here we conduct a genetic inventory of essentially all laboratory stocks of the Mangrove Killifish held worldwide. At 32 microsatellite loci, these stocks proved to show extensive interline differentiation as well as some intraline variation, much of which can be attributed to post-origin *de novo* mutations and/or to the segregation of polymorphisms from wild progenitors. Our genetic findings also document that many of the surveyed laboratory strains are not what they have been labeled, apparently due to the rather frequent mishandling or unintended mixing of various laboratory stocks over the years. Our genetic inventory should help to clarify much of this confusion about the clonal identities and genetic relationships of laboratory lines, and thereby help to rejuvenate interest in *K. marmoratus* as a reliable vertebrate model for experimental research that requires or can capitalize upon "clonal" replicate specimens.

ADDENDUM

JCA's long-term experience with hermaphroditic Kmar provided a motivation for his 2011 book entitled "Hermaphroditism: A Primer on the Biology, Ecology, and Evolution of Dual Sexuality."

Extreme homogeneity and low genetic diversity in *Kryptolebias ocellatus* from southeastern Brazil suggest a recent foundation for this androdioecious fish population

Tatarenkov, A., S.M.Q. Lima, and J.C. Avise. 2011. Extreme homogeneity and low genetic diversity in Kryptolebias ocellatus *from southeastern Brazil suggest a recent foundation for this androdioecious fish population.* Journal of Fish Biology 79:2095–2105.

ANECDOTE OR BACKDROP

This study continued to address the evolutionary origins of the Kmar clade.

ABSTRACT

This study documents unexpectedly low levels of intra- and interpopulation genetic diversity in *Kryptolebias ocellatus*, an androdioecious and predominantly self-fertilizing killifish from southeastern Brazil. This finding generally is inconsistent with established opinion that the *K. ocellatus/marmoratus* clade originated in this geographic region and later dispersed northward into the Caribbean.

Microevolutionary distribution of isogenicity in a self-fertilizing fish (*Kryptolebias marmoratus*) in the Florida Keys

*Tatarenkov, A., Earley, R.L, Taylor D.S., and Avise, J.C. 2012. Microevolutionary distribution of isogenicity in a self-fertilizing fish (*Kryptolebias marmoratus*) in the Florida Keys.* Integrative and Comparative Biology 52:743–752.

ANECDOTE OR BACKDROP

This study began to address ecological and behavioral factors that might underlie Kmar's unique propensity to self-fertilize.

ABSTRACT

A survey of 32 highly polymorphic loci in >200 specimens of a self-fertilizing fish (*Kryptolebias marmoratus*) from multiple locales in the Florida Keys revealed extensive population genetic structure on microspatial and microtemporal scales. Observed heterozygosities were severely constrained, as expected for a hermaphroditic species with a mixed-mating system and low rates of outcrossing. Despite the pronounced population structure and the implied restrictions on effective gene flow, isogenicity (genetic identity across individuals) within and among local inbred demes was surprisingly low even after factoring out probable *de novo* mutations. Results can be interpreted as inconsistent with the notion that genetic adaptation to local environmental conditions is the primary driving force impacting multilocus population genetic architecture in this self-fertilizing vertebrate species.

Allard's argument versus Baker's contention for the adaptive significance of selfing in a hermaphroditic fish

Avise J.C. and A. Tatarenkov. 2012. Allard's argument versus Baker's contention for the adaptive significance of selfing in a hermaphroditic fish. **Proceedings of the National Academy of Sciences USA** *109:18862–18867.*

ANECDOTE OR BACKDROP

This study continued our attempt to interpret the evolutionary significance of self-fertilization in the Kmar clade. Results suggest that fertilization assurance is probably a key selective factor favoring the ability of each Kmar individual to reproduce without the need to find and court a mate.

ABSTRACT

Fertilization assurance (Baker's contention) and multilocus coadaptation (Allard's argument) are two distinct hypotheses for the adaptive significance of self-fertilization in hermaphroditic taxa, with both scenarios having been invoked to rationalize isogenicity via incest in various plants and invertebrate animals with predominant selfing. Here we contrast Allard's argument and Baker's contention as applied to the world's only known vertebrate that routinely self-fertilizes. We pay special attention to frequencies of locally most common multilocus genotypes (LMCMLGs) in Floridian populations of the Mangrove Rivulus (*Kryptolebias marmoratus*). Isogenicity patterns in this fish appear inconsistent with Allard's argument, thus by default leaving Baker's contention as the more plausible scenario (a result also supported by natural-history information for this species). These results contrast with the isogenicity patterns and conclusions previously drawn from several self-fertilizing plants and invertebrate animal species. Thus the adaptive significance of selfing apparently varies across hermaphroditic taxa.

ADDENDUM

Fertilization assurance is sometimes also referred to as fertilization insurance, because being both male and female at the same time is somewhat like having an insurance policy against reproductive failure due to any inability to secure a mate.

Hundreds of SNPs versus dozens of SSRs: which dataset better characterizes natural clonal lineages in a self-fertilizing fish?

Mesak, F., A. Tatarenkov, R.L. Earley, and J.C. Avise. 2014. Hundreds of SNPs versus dozens of SSRs: which dataset better characterizes natural clonal lineages in a self-fertilizing fish? Frontiers in Ecology and Evolution 2 (article 74):1–8.

ANECDOTE OR BACKDROP

In recent years, "next-generation" sequencing has promised to revolutionize the fields of population genetics and evolution by providing ready access to voluminous amounts of genetic information from essentially any species. In this paper, we examine to what extent (if any) data from one next-generation procedure might outperform traditional data from highly polymorphic microsatellite loci. The results of this study highlight some of the pitfalls as well as the promises of next-generation single-nucleotide polymorphisms (SNPs) from a nonmodel species for which a fully sequenced reference genome is unavailable.

ABSTRACT

For more than two decades, mitochondrial DNA sequences and simple sequence repeats (SSRs or microsatellite loci) have served as gold standards in population genetics. More recently, next-generation sequencing (NGS) has enabled researchers to address biological questions that can benefit from hundreds or even thousands of nuclear single-nucleotide polymorphisms (SNPs) generated by restriction-site associated DNA sequencing (RAD-seq). Here we compare the performance of SSR and SNP methods to characterize clonal patterns in a self-fertilizing and highly inbred killifish, *Kryptolebias marmoratus* (mangrove rivulus) in Florida. RAD-seq analyses conducted on 18 inbred lineages of mangrove rivulus obtained from western Florida and a distant location in eastern Florida unveiled 481 polymorphic Rad loci of which 129 were homozygous within specimens and 352 loci were heterozygous in at least one individual. An initial UPGMA phenogram was constructed, based on 32 microsatellite loci, and used as a benchmark for comparisons with SNP-based phenograms, using a number of different criteria for SNP selection. A phenogram produced by the homozygous SNPs was in excellent agreement with the one generated from 32 microsatellite loci. Heterozygous SNP data and Rad loci with more than one polymorphic site contributed more noise than usable signal and were unable to resolve clades consistently. This is likely due to errors in identifying homologous loci in the absence of a reference genome. In summary, although the RAD data were powerful in distinguishing the clonal lineages identified by SSR analyses, they also carried considerable phylogenetic noise. Our results suggest that RAD-seq methods should be used with caution for inferring fine population structure, and that stringent quality controls are necessary to reduce false phylogenetic signals.

ADDENDUM

Despite it evident resolving power, next-generation sequencing does have some serious limitations that need to be recognized when applying this approach to evolutionary questions for nonmodel taxa.

Amphibians

INTRODUCTION

Among the various taxonomic classes of vertebrates, Amphibia has been the least studied in JCA's laboratory. Nevertheless, a few genetic studies on amphibians have emerged, mostly spearheaded by Trip Lamb, a superb herpetologist who was a graduate student in the lab in the early 1980s, right at the dawn of the era when female-transmitted mitochondrial DNA was first being employed in population-level studies. Although only a few "data papers" emerged from this work on amphibians, the genetic findings on a hybrid population of tree frogs did provide the primary motivation for the collaborative development and deployment of cytonuclear disequilibrium statistics by the theoreticians Marjorie Asmussen and Jonathan Arnold. This research exemplifies how molecular data can both motivate and be informed by population genetic theory.

Directional introgression of mitochondrial DNA in a hybrid population of tree frogs: the influence of mating behavior

Lamb, T. and J.C. Avise. 1986. Directional introgression of mitochondrial DNA in a hybrid population of tree frogs: the influence of mating behavior. **Proceedings of the National Academy of Sciences USA 83:2526–2530.**

ANECDOTE OR BACKDROP

Trip Lamb was a highly knowledgable herpetologist long before he joined JCA's laboratory, having been enamored of amphibians and reptiles ever since his childhood. Thus, as soon as he learned some molecular genetic techniques in graduate school, he had a long list of creatures and project settings where they could be meaningfully applied. This next study is a case-in-point. In it, Trip genetically revisited a long-known hybrid population of tree frogs near Auburn, AL. His cytonuclear dissection of this hybrid population using mitochondrial and nuclear markers revealed the remarkable extent to which the distinctive mating behaviors of related species can shape the genetic architecture of a hybrid swarm.

ABSTRACT

A total of 305 individuals from a hybrid population of North American tree frogs were characterized for allozyme and mitochondrial (mt) DNA genotype. Species-specific mating behaviors had suggested the potential for directional hybridization, in which matings between *Hyla cinerea* males and *Hyla gratiosa* females numerically predominate over the reciprocal combination. Such directional bias leads to predictions about expected distributions of the female-transmitted mtDNA markers in F_1, backcross, and later-generation hybrids. These predictions were fully confirmed by the observed distributions of mtDNA genotypes among these allozymically inferred hybrid classes. Results exemplify the significance of stereotyped mating behaviors in determining the genetic architecture of a hybrid population.

ADDENDUM

This early study provided one of the most comprehensive cytonuclear dissections of a hybrid zone available at that time for any species.

FIGURE 8.1 Green Tree Frog, *Hyla cinerea*.

Morphological variability in genetically defined categories of anuran hybrids

Lamb, T. and J.C. Avise. 1987. Morphological variability in genetically defined categories of anuran hybrids. Evolution 41:157–165.

ANECDOTE OR BACKDROP

This was a follow-up paper in which Trip Lamb compared his genetic inferences about the hybrid tree frog population against inferences that would have come from morphology-based appraisals alone. The results of this comparison make a strong case for the exceptional power of molecular markers in hybrid settings.

ABSTRACT

Hybridization phenomena in anurans have traditionally been studied through morphological comparisons, under the assumption that various hybrids (e.g., F_1's, backcrosses) are predictably intermediate to parental species. We critically evaluate this assumption by examining morphology in genetically categorized hybrids between the tree frogs *H. cinerea* and *H. gratiosa*. A total of 202 frogs from a hybridizing population in Alabama were assayed for allozyme and mitochondrial DNA genotype and for a large suite of osteological characters. Discriminant analyses demonstrated distinct morphological separation between the genetically "pure" parental species. Morphometric analyses of genetically identified hybrids showed: (i) an extreme range of phenotypic expression within F_1 and backcross classes and (ii) no apparent directional parental bias on the F_1 phenotype. Had morphology alone been used as a guide, over 40 percent of the individuals with known hybrid ancestry would have been misclassified as "pure" parental species, and about 25 percent of the backcross individuals would not have been distinguished from F_1's. These results exemplify the utility of joint comparisons of morphology and genotypic constitution in studies of natural hybridization, and they emphasize the limitations inherent in describing hybrid classes solely by morphological criteria.

ADDENDUM

Thanks in no small part to this article, it is now widely appreciated that genetic appraisals should accompany morphological appraisals in almost any natural hybrid setting.

Definition and properties of disequilibrium statistics for associations between nuclear and cytoplasmic genotypes

Asmussen, M.A., J. Arnold, and J.C. Avise. 1987. Definition and properties of disequilibrium statistics for associations between nuclear and cytoplasmic genotypes. Genetics 115:755–768.

ANECDOTE OR BACKDROP

This study basically introduced the notion of cytonuclear disequilibria (nonrandom associations between nuclear and cytoplasmic [such as mitochondrial] genotypes). It reflects work that was stimulated in part by Trip Lamb's data on hybridizing tree frogs, and was done in collaboration with theoreticians Marjorie Asmussen and Jonathan Arnold (two other faculty members at the University of Georgia).

ABSTRACT

We define and establish the interrelationships of four components of statistical association between a diploid nuclear gene and a uniparentally transmitted, haploid cytoplasmic gene: an allelic (gametic) disequilibrium (D), which measures associations between alleles at the two loci; and three genotypic disequilibria (D_1, D_2, D_3), which measure associations between two cytotypes and the three respective nuclear backgrounds. We also consider an alternative set of measures, including D and the residual disequilibrium (d). The dynamics of these disequilibria are then examined under three conventional models of the mating system: (1) random mating; (2) assortative mating without dominance (the "mixed-mating model"); and (2b) assortative mating with dominance. The trajectories of gametic disequilibria are similar to those for pairs of unlinked nuclear loci. The dynamics of genotypic disequilibria exhibit a variety of behaviors depending on the model and the initial conditions. Procedures for statistical estimation of cytonuclear disequilibria are developed and applied to several real and hypothetical data sets. Special attention is paid to the biological interpretations of various categories of allelic and genotypic disequilibria in hybrid zones. Genetic systems for which these statistics might be appropriate include nuclear genotype frequencies in conjunction with those for mitochondrial DNA, chloroplast DNA, or cytoplasmically inherited microorganisms.

ADDENDUM

Although statistics for gametic-phase disequilibria among nuclear genes were available, this was essentially the first serious attempt to develop analogous statistics for nuclear and cytoplasmic genes considered jointly.

An epistatic mating system model can produce permanent cytonuclear disequilibria in a hybrid zone

Arnold, J., M.A. Asmussen, and J.C. Avise. 1988. An epistatic mating system model can produce permanent cytonuclear disequilibria in a hybrid zone. Proceedings of the National Academy of Sciences USA 85:1893–1896.

ANECDOTE OR BACKDROP

This and the paper that follow are the second and third in a series of theoretical articles on the evolutionary dynamics of cytonuclear disequilibria under various hypothetical population genetic scenarios. From this collaboration, JCA came to appreciate more fully that different peoples' minds really do operate in very different fashions. To Marjorie Asmussen, a biological phenomenon could not be genuinely understood unless and until it was formalized in the guise of mathematical equations; whereas for JCA no amount of mathematicization could make a phenomenon truly understandable unless it could also be intuited more directly from biological observations.

ABSTRACT

We examine the evolutionary dynamics of gametic and genotypic disequilibria between a cytoplasmic gene and a nuclear gene under two mating system models relevant to hybrid zones. In the first model, in which female mating preference is determined by an epistatic interaction between the two loci, permanent nonzero cytonuclear disequilibria are possible for a variety of initial genotype frequencies, particularly when rates of assortative mating for the two parental species are high. In contrast, when mating preference is effectively determined by interaction between a cytoplasmic gene and the multilocus genotype characteristic of the parental species, all cytonuclear disequilibria, as well as frequencies of pure parentals, rapidly decay to zero unless assortative mating is nearly perfect. Results of the models are applied to the interpretation of observed cytonuclear associations in a hybrid population of *Hyla* tree frogs.

The effects of assortative mating and migration on cytonuclear associations in hybrid zones

Asmussen, M., J. Arnold, and J.C. Avise. 1989. The effects of assortative mating and migration on cytonuclear associations in hybrid zones. Genetics 122:923–934.

ABSTRACT

We examine the influence of nonrandom mating and immigration on the evolutionary dynamics of cytonuclear associations in hybrid zones. Recursion equations for allelic and genotypic cytonuclear disequilibria were generated under models of: (i) migration alone, assuming hybrid zone matings are random with respect to cytonuclear genotype and (ii) migration in conjunction with refined epistatic mating, in which females of the pure parental species preferentially mate with conspecific males. Major results are as follows: (i) even the slightest migration removes the dependency of the final outcome on initial conditions, producing a unique equilibrium in which both pure parental genotypes are maintained in the hybrid zone; (ii) in contrast to nuclear genes, the dynamics of cytoplasmic allele frequencies appear robust to changes in the assumed mating system, yet are particularly sensitive to gene flow; (iii) continued immigration can generate permanent cytonuclear disequilibria, whether mating is random or assortative; and (iv) the order of population censusing (before vs. after reproduction by immigrants) can have a dramatic effect on the magnitude but not the pattern of cytonuclear disequilibria. Using the maximum likelihood method, the parameter space of migration rates and assortative mating rates was examined for best fit to observed cytonuclear disequilibria data in a hybrid population of *Hyla* tree frogs. An epistatic mating model with a total immigration rate of about 32% per generation produces equilibrium gene frequencies and cytonuclear disequilibria consistent with the empirical observations.

ADDENDUM

Marjorie Asmussen died tragically in 2004, just as she was beginning to hit full stride in her follow-ups to this and other theoretical work on nuclear-cytoplasmic interactions.

Size polymorphism and heteroplasmy in the mitochondrial DNA of lower vertebrates

Bermingham, E., T. Lamb, and J.C. Avise. 1986. Size polymorphism and heteroplasmy in the mitochondrial DNA of lower vertebrates. **Journal of Heredity 77:249–252.**

ANECDOTE OR BACKDROP

Mitochondrial DNA introduced a new level to the hierarchy of populations that evolutionary geneticists must understand. Whereas a typical nuclear gene occurs in only two copies per diploid cell and one copy per gamete, large populations of mtDNA molecules co-inhabit each somatic or germline cell. Thus, mitochondrial geneticists must concern themselves not only with populations of organisms but also with populations of mtDNA molecules inside each individual. Early in the mtDNA era, JCA and his colleagues began to examine such intraindividual population phenomena from both theoretical and empirical vantages. The next paper offers an example of the latter.

ABSTRACT

The mitochondrial DNA of the bowfin fish and each of two species of tree frogs displays large-scale size variation. Within each species, mitochondrial genomes span more than a 700 base pair range, and the size polymorphism is localized to one portion of the genome. In addition, about 5 percent of the total 357 individuals surveyed were observed to carry two size classes of mtDNA. These findings are among the few documented instances of extensive within-species mtDNA size polymorphism and individual heteroplasmy, and constitute exceptions to previously reached generalizations about the molecular basis of mtDNA variation.

ADDENDUM

Although the phenomena of mtDNA size polymorphism and heteroplasmy (the presence of two or more mtDNA genotypes within an individual) have now been reported for numerous vertebrate and invertebrate species, they are normally quite circumscribed and therefore do not seriously compromise phylogeographic or other population genetic analyses.

CHAPTER

9

Marine Turtles

INTRODUCTION

Seven extant species of turtle inhabit the world's oceans. Essentially all of these ancient mariners are endangered or of special conservation concern. When Christopher Columbus sailed into the Caribbean Sea during his voyage of discovery, he noted in his logbook how members of his crew were disquieted by the ship's many collisions with sea turtles, which clearly were abundant at that time. Sadly, turtle numbers today are greatly diminished mostly due to human impacts such as overharvesting of adults and eggs for food, and the loss or disturbance of many turtle nesting beaches around the world.

Traditionally, studies of sea turtles were based on behavioral field observations (which are notoriously difficult when turtles are at sea), morphological appraisals, and physical tagging studies to monitor individual movements. Such analyses nevertheless left many knowledge gaps that begged for genetic reappraisals. Beginning in the late 1980s, two graduate students (Brian Bowen and Steve Karl) in JCA's laboratory spearheaded the first such molecular genetic analyses of various marine turtle species. Their pioneering research eventuated in many novel scientific findings, including those summarized in this chapter.

An odyssey of the green sea turtle: Ascension Island revisited

Bowen, B.W., A.B. Meylan, and J.C. Avise. 1989. An odyssey of the green sea turtle: Ascension Island revisited. **Proceedings of the National Academy of Sciences USA 86:573–576.**

ANECDOTE OR BACKDROP

Ascension Island sits atop the mid-Atlantic ridge (halfway between South America and Africa) and is one of the most isolated pieces of land on Earth, yet it is also home to one of the major nesting rookeries for green turtles in the whole Atlantic basin. This begs the question: how and when did Green Turtles arrive on Ascension? This paper helped to provide the answers. It required that JCA and his students travel to Ascension (by U.S. Air Force jet, and with formal permission from the Assistant Secretary of Defense) to collect the turtle eggs used in the genetic analyses. This 2-week collecting expedition was definitely one of the most memorable experiences of JCA's life.

ABSTRACT

Green turtles (*Chelonia mydas*) that nest on Ascension Island, in the south-central Atlantic, utilize feeding grounds along the coast of Brazil, more than 2,000 kilometers away. To account for the origins of this remarkable migratory behavior, Carr and Coleman (*Nature* 249:128–130) proposed a vicariant biogeographic scenario involving plate tectonics and natal homing. Under the Carr–Coleman hypothesis, the ancestors of Ascension Island green turtles nested on islands adjacent to South America in the late Cretaceous, soon after the opening of the equatorial Atlantic Ocean. Over the last 70 million years, these volcanic islands have been displaced from South America by sea-floor spreading, at a rate of about 2 centimeter per year. A population-specific instinct to migrate to Ascension Island is thus proposed to have evolved gradually over tens of millions of years of genetic isolation. Here we critically test the Carr–Coleman hypothesis by assaying genetic divergence among several widely separated green turtle rookeries. We have found fixed or nearly fixed mitochondrial (mt) DNA restriction site differences between some Atlantic rookeries, suggesting a severe restriction on contemporary gene flow. Data are consistent with a natal homing hypothesis. However, an extremely close similarity in overall mtDNA sequences of surveyed Atlantic green turtles from three rookeries is incompatible with the Carr–Coleman scenario. The colonization of Ascension Island, or at least extensive gene flow into the population, has been evolutionarily recent.

ADDENDUM

Ascension Island has no native peoples, but it does house U.S. and British Air Force bases. It also had been a major nesting site for many seabirds, before introduced cats and rats decimated its avian populations.

FIGURE 9.1   Green turtle, *Chelonia mydas.*

A genetic test of natal homing versus social facilitation models for green turtle migration

Meylan, A.B., B.W. Bowen, and J.C. Avise. 1990. A genetic test of natal homing versus social facilitation models for green turtle migration. Science 248:724–727.

ANECDOTE OR BACKDROP

Brian Bowen was a brilliant and ambitious graduate student who spearheaded most of the marine turtle research in JCA's laboratory during the late 1980s and early 1990s. In collaboration with turtle expert Anne Meylan, Brian conceived a powerful way to decipher whether sea turtles typically return to their natal sites to nest. The empirical test involved analyses of mitochondrial DNA sequences from multiple rookeries, the results of which are summarized in the paper from which the following abstract was taken. Today it seems to be standard lore among turtle biologists that gravid females often natally home, but that conclusion had never been empirically confirmed before this pathbreaking genetic study was published.

ABSTRACT

Female green turtles exhibit strong nest-site fidelity, but whether the nesting beach is the natal site is not known. Under the natal homing hypothesis, females return to their natal beach to nest, whereas under the social facilitation model, virgin females follow experienced breeders to nesting beaches and after a "favorable" nesting experience, fix on that site for future nestings. Differences shown in mitochondrial DNA genotype frequency among green turtle colonies in the Caribbean Sea and Atlantic Ocean are consistent with natal homing expectations and indicate that social facilitation to nonnatal sites is rare.

ADDENDUM

Prior to this genetic study, essentially all studies of sea turtle migration had relied upon returns of physical tags attached to nesting females. Although such physical tagging experiments had revealed that adult females are site faithful in their nesting behaviors, they could not address intergenerational or past migrational patterns.

Evolutionary distinctiveness of the endangered Kemp's Ridley sea turtle

Bowen, B.W., A.B. Meylan, and J.C. Avise. 1991. Evolutionary distinctiveness of the endangered Kemp's Ridley sea turtle. Nature 352:709–711.

ANECDOTE OR BACKDROP

In the cases of the Dusky Seaside Sparrow (see Chapter 12) and the Colonial Pocket Gopher (see Chapter 13), genetic appraisals in JCA's lab called into question conventional taxonomic assignments for these endangered species. However, molecular reappraisals of other taxonomically suspect species might in some cases bolster the evolutionary rationale for conservation efforts. A case-in-point from JCA's laboratory involved the Kemp's Ridley turtle, which nests at a single locale in the western Gulf of Mexico and has been the subject of the largest international preservation effort for any marine turtle. But the Kemp's Ridley has a close evolutionary relative—the Olive Ridley—that is widely distributed and relatively abundant. Just how close, phylogenetically, are the Olive Ridley and Kemp's Ridley, and does the latter merit special conservation attention? This article answered such questions.

ABSTRACT

The endangered Kemp's Ridley sea turtle (*Lepidochelys kempi*) nests almost exclusively at a single locality in the western Gulf of Mexico, whereas the olive ridley (*L. olivacea*) nests globally in warm oceans. Morphological similarities between *kempi* and *olivacea*, and a geographic distribution that makes no sense at all under current conditions of climate and geography, raise questions about the degree of evolutionary divergence between these taxa. Analysis of mitochondrial (mt) DNA restriction sites shows that Kemp's ridley is distinct from the olive ridley in matriarchal phylogeny, and that the two are sister taxa with respect to other marine turtles. Separation of olive and the Kemp's ridley lineages may date to formation of the Isthmus of Panama, whereas the global spread of the olive ridley lineage occurred recently. In contrast to some other recent examples in which molecular genetic assessments challenged systematic assignments underlying conservation programs, our mtDNA data corroborate the taxonomy of an endangered form.

ADDENDUM

The Kemp's Ridley remains the rarest of all marine turtle species and hence continues to be a critically endangered species of special conservation concern.

FIGURE 9.2   Kemp's Ridley Turtle, *Lepidochelys kempi*.

Mitochondrial DNA evolution at a turtle's pace: evidence for low genetic variability and reduced microevolutionary rate in the Testudines

Avise, J.C., B.W. Bowen, T. Lamb, A.B. Meylan, and E. Bermingham. 1992. Mitochondrial DNA evolution at a turtle's pace: evidence for low genetic variability and reduced microevolutionary rate in the Testudines. Molecular Biology and Evolution 9:457–473.

ANECDOTE OR BACKDROP

By 1990, it had become standard dogma that animal mtDNA evolves very rapidly at the nucleotide sequence level (perhaps $5 \times -10 \times$ faster that typical single-copy nuclear DNA). Thus, it came as some surprise when we uncovered consistent evidence—from several turtle taxa—for a significant deceleration in mtDNA rate relative to the vertebrate norm. Although sufficient intraspecific mtDNA variation for phylogeographic analyses remained present, the multifaceted evidence summarized in this report suggested that mtDNA evolutionary clocks in the testudines warrant some recalibration.

ABSTRACT

Evidence is compiled suggesting a slowdown in mean microevolutionary rate for turtle mitochondrial (mt) DNA. Within each of six species or species complexes of Testudines, representing six genera and three taxonomic families, sequence divergence estimates derived from restriction assays are consistently lower than expectations based on either (i) the dates of particular geographic barriers with which significant mtDNA genetic clades appear associated or (ii) the magnitudes of sequence divergence between mtDNA clades in nonturtle species that otherwise exhibit striking phylogeographic concordance with the genetic partitions in turtles. Magnitudes of the inferred rate slowdowns average eightfold relative to the "conventional" mtDNA clock calibration of 2%/Myr sequence divergence between higher animal lineages. Reasons for the postulated deceleration remain unknown, but two intriguing correlates are (i) the exceptionally long generation length of most turtles and (ii) turtles' low metabolic rate. Both factors have been suspected of influencing evolutionary rates in the DNA sequences of some other vertebrate groups. Uncertainties about the dates of cladogenetic events in these Testudines leave room for alternatives to the slowdown interpretation, but consistency in the direction of the inferred pattern, across several turtle species and evolutionary settings, suggests the need for caution in acceptance of a universal mtDNA clock calibration for higher animals.

ADDENDUM

Thus study was part of the then-growing realization that any "molecular clock" for vertebrate taxa must be far from universal.

Global population structure and natural history of the green turtle (*Chelonia mydas*) in terms of matriarchal phylogeny

Bowen, B.W., A.B. Meylan, J. Perran Ross, C.J. Limpus, G.H. Balazs, and J.C. Avise. 1992. Global population structure and natural history of the green turtle (Chelonia mydas) in terms of matriarchal phylogeny. Evolution 46:865–881.

ANECDOTE OR BACKDROP

This genetic study of global phylogeography of the green turtle might seem deceptively simple, unless the extraordinary logistical difficulties of getting the tissue samples from around the world are taken into full account. For each of the many collections, we not only had to arrange for the turtle specimens to be collected and shipped to the laboratory (on buffer or ice), but we also had to have in hand the many necessary collecting permits from the sampled countries, the recipient country (the United States), and any international authorities that were also involved (such as CITES—the Convention on International Trade in Endangered Species). We sometimes joked (although it was quite genuinely true) that Brian Bowen spent at least 90% of his time arranging for the collecting permits and travel logistics, and only 10% of his time actually gathering and analyzing the genetic data in the lab and at the computer.

ABSTRACT

To address aspects of the evolution and natural history of Green Turtles, we assayed mitochondrial (mt) DNA genotypes from 226 specimens representing 15 major rookeries around the world. Phylogenetic analyses of these data revealed (i) a comparatively low level of mtDNA variability and a slow mtDNA evolutionary rate (relative to estimates for many other vertebrates); (ii) a fundamental phylogenetic split distinguishing all green turtles in the Atlantic—Mediterranean from those in the Indian—Pacific Oceans; (iii) no evidence for matrilineal distinctiveness of a commonly recognized taxonomic form in the East Pacific (the black turtle *C.m. agassizi* or *C. agassizi*); (iv) in opposition to published hypotheses, a recent origin for the Ascension Island rookery, and its close genetic relationship to a geographically proximate rookery in Brazil; and (v) a geographic population substructure within each ocean basin (typically involving fixed or nearly fixed genotypic differences between nesting populations) that suggests a strong propensity for natal homing by females. Overall, the global matriarchal phylogeny of *Chelonia mydas* appears to have been shaped by both geography (ocean basin separations) and behavior (natal homing on regional or rookery-specific scales). The shallow evolutionary population structure within ocean basins likely results from demographic turnover (extinction and colonization) of rookeries over timeframes that are short by evolutionary standards but long by ecological standards.

ADDENDUM

Although this study involved mtDNA only, it was of special interest because of the close linkages among the following: female population demography, the genealogical and migrational histories of female turtles, and conservation efforts.

Global population genetic structure and male-mediated gene flow in the green turtle (*Chelonia mydas*): RFLP analyses of anonymous nuclear loci

*Karl, S.A., B.W. Bowen, and J.C. Avise. 1992. Global population genetic structure and male-mediated gene flow in the green turtle (*Chelonia mydas): RFLP analyses of anonymous nuclear loci. Genetics 131:163–173.*

ANECDOTE OR BACKDROP

Steve Karl was another brilliant and hard-working student of JCA that early on also became fully engaged in the sea turtle projects. Until that point in time, we had relied primarily on mtDNA to draw our inferences about marine turtle behavior and biology. Steve's first task, as summarized in this report, was to extend such population genetic studies to the nuclear genome, such that inferences could be made about the migrational behaviors of males (as well as females).

ABSTRACT

We introduce an approach for the analysis of Mendelian polymorphisms in nuclear DNA (nDNA), using restriction fragment patterns from anonymous single-copy regions amplified by the polymerase chain reaction, and apply this method to the elucidation of population structure and gene flow in the endangered green turtle, *Chelonia mydas*. Seven anonymous clones isolated from a total cell DNA library were sequenced to generate primers for the amplification of nDNA fragments. Nine individuals were screened for restriction site polymorphisms at these 7 loci, using 40 endonucleases. Two loci were monomorphic, while the remainder exhibited a total of nine polymorphic restriction sites and three size variants (reflecting 600-basepair (bp) and 20-bp deletions and a 20-bp insertion). A total of 256 turtle specimens from 15 nesting populations worldwide were then scored for these polymorphisms. Genotypic proportions within populations were in accord with Hardy–Weinberg expectations. Strong linkage disequilibrium observed among polymorphic sites within loci enabled multisite haplotype assignments. Estimates of the standardized variance in haplotype frequency among global collections ($F_{ST} = 0.17$), within the Atlantic–Mediterranean ($F_{ST} = 0.13$), and within the Indian–Pacific ($F_{ST} = 0.13$), revealed a moderate degree of population substructure. Although a previous study concluded that nesting populations appear to be highly structured with respect to female (mitochondrial DNA) lineages, estimates of *Nm* based on nDNA from this study indicate moderate rates of male-mediated gene flow. A positive relationship between genetic similarity and geographic proximity suggests historical connections and/or contemporary gene flow between particular rookery populations, likely via matings on overlapping feeding grounds, migration corridors, or nonnatal rookeries.

ADDENDUM

The RFLP approaches introduced in this article have largely been superseded in recent years by more direct approaches for sequencing all or parts of the nuclear genome.

Population structure of loggerhead turtles (*Caretta caretta*) in the northwestern Atlantic Ocean and Mediterranean Sea

Bowen, B.W., J.C. Avise, J.I. Richardson, A.B. Meylan, D. Margaritoulis, and S.R. Hopkins-Murphy. 1993. Population structure of loggerhead turtles (Caretta caretta) *in the northwestern Atlantic Ocean and Mediterranean Sea.* **Conservation Biology 7:834–844.**

ANECDOTE OR BACKDROP

This was the first phylogeographic study of yet another marine turtle—the Loggerhead. It was a prelude to subsequent analyses that would extend the genetic surveys of this species to a truly global scale (see the next abstract). Again, a major part of these efforts was merely getting the samples to begin with.

ABSTRACT

To assess population genetic structure and evolutionary relationships among nesting populations of loggerhead turtles (*Caretta caretta*), we analyzed mitochondrial (mt) DNA variation in 113 samples from four nesting beaches in the northwestern Atlantic Ocean and from one nesting beach in the Mediterranean Sea. Significant differences in haplotype frequency between nesting populations in Florida and in Georgia/South Carolina, and between both of these assemblages and the Mediterranean rookery, indicate substantial restrictions on contemporary gene flow between regional populations, and therefore a strong tendency for natal homing by females. Nonetheless, this regional genetic structure appears shallow, indicating recent evolutionary connections among rookeries. Data from tag returns and mtDNA, as well as geological considerations, suggest that over short evolutionary timescales (perhaps a few thousand years), dispersal by female loggerheads is sufficient to allow colonization of appropriate habitat in proximity to established rookeries, but is too low to significantly affect the population dynamics of rookeries on a contemporary timescale. These data indicate that nesting populations of the loggerhead turtle must be managed as demographically independent units. The population subdivisions based on mtDNA analyses are concordant with previously reported distinctions between Florida and Georgia/South Carolina nesting populations based on environmental markers, tag recaptures, and morphology.

FIGURE 9.3   Loggerhead turtle, *Caretta caretta*.

Global phylogeography of the loggerhead turtle (*Caretta caretta*) as indicated by mitochondrial DNA haplotypes

Bowen, B.W., N. Kamezaki, C.J. Limpus, G.R. Hughes, A.B. Meylan, and J.C. Avise. 1994. Global phylogeography of the loggerhead turtle (Caretta caretta) *as indicated by mitochondrial DNA haplotypes.* Evolution *48:1820–1828.*

ANECDOTE OR BACKDROP

This study of global phylogeographic structure not only illuminated the history of connections among scattered rookeries within this species but also permitted direct comparisons against our similar phylogeographic analyses previously conducted on the green turtle.

ABSTRACT

Restriction-site analyses of mitochondrial (mt) DNA from the loggerhead sea turtle (*Caretta caretta*) reveal substantial phylogeographic structure among major nesting populations in the Atlantic, Indian, and Pacific Oceans and the Mediterranean Sea. Based on 176 samples from 8 nesting populations, most breeding colonies were distinguished from other assayed nesting locations by diagnostic and often fixed restriction-site differences, indicating a strong propensity for natal homing by nesting females. Phylogenetic analyses revealed two distinctive matrilines in the loggerhead turtle that differ by a mean estimated sequence divergence $p = 0.009$, a value similar in magnitude to the deepest intraspecific mtDNA node ($p = 0.007$) reported in a global survey of the green sea turtle *Chelonia mydas*. In contrast to the green turtle, where a fundamental phylogenetic split distinguished turtles in the Atlantic Ocean and the Mediterranean Sea from those in the Indian and Pacific Oceans, genotypes representing the two primary loggerhead mtDNA lineages were observed in both Atlantic–Mediterranean and Indian–Pacific samples. We attribute this aspect of phylogeographic structure in *Caretta caretta* to recent interoceanic gene flow, probably mediated by the ability of this temperate-adapted species to utilize habitats around southern Africa. These results demonstrate how differences in the ecology and geographic ranges of marine turtle species can influence their comparative global population structures.

A molecular phylogeny for marine turtles: trait mapping, rate assessment, and conservation relevance

Bowen, B.W., W.S. Nelson, and J.C. Avise. 1993. A molecular phylogeny for marine turtles: trait mapping, rate assessment, and conservation relevance. Proceedings of the National Academy of Sciences USA *90:5574–5577.*

ANECDOTE OR BACKDROP

A powerful technique in comparative evolution involves mapping phenotypic traits onto phylogenetic trees that have been erected using independent molecular data. With the phylogeny of a taxonomic group providing the historical backdrop, researchers can map alternative phenotypic conditions onto the tree and thereby reconstruct the most probable course of evolutionary interconversions among them. This approach is sometimes called "phylogenetic character mapping." This paper on sea turtle phylogeny provides one empirical example from JCA's laboratory (see Chapter 16 for other such examples).

ABSTRACT

Nucleotide sequences from the cytochrome *b* gene of mitochondrial DNA were employed to resolve phylogenetic controversies and to assess molecular evolutionary rates in marine turtles (Chelonioidea). Findings of special relevance to conservation biology include discovery of a distant relationship between *Natator* and other cheloniid species, the paraphyly of *Chelonia mydas* with respect to *Chelonia agassizi,* and genetic distinctiveness of *Lepidochelys kempi* from *Lepidochelys olivacea.* A longstanding debate in evolutionary ecology was resolved by phylogenetic mapping of dietary habits, which indicates that the spongivore *Eretmochelys imbricata* evolved from a carnivorous rather than an herbivorous ancestor. Sequence divergences at intergeneric and interfamilial levels, when assessed against fossil-based separation times, support previous suggestions (from microevolutionary comparisons) that mitochondrial DNA in marine turtles evolves much more slowly than under the "conventional" vertebrate clock. This slow pace of nucleotide replacement is consistent with recent hypotheses linking substitution rate to generate length and metabolic pace.

ADDENDUM

Many more examples of phylogenetic character mapping for a wide range of creatures are highlighted in JCA's (2006) book entitled "Evolutionary Pathways in Nature: A Phylogenetic Approach."

Investigating sea turtle migration using DNA markers

Avise, J.C. and B.W. Bowen. 1994. Investigating sea turtle migration using DNA markers. Current Opinions in Genetics and Development 4:882–886.

ANECDOTE OR BACKDROP

This brief abstract is from one of several review articles in which we summarized our cumulative molecular genetic experiences with all of the marine turtle species.

ABSTRACT

The past year has seen a further marshaling of genetic evidence for natal homing in several species of marine turtles, a phenomenon wherein females, upon reaching sexual maturity, exhibit a propensity to return to nest at or near the respective beaches upon which they hatched some two or more decades earlier. This genetics-based inference stems from the strong spatial patterning observed in mitochondrial DNA lineages among nesting sites. Rookery-specific mitochondrial DNA markers are now being employed to monitor the natal sources of individuals captured at other phases of the life cycle, and the genetic findings have important conservation ramifications.

Hybridization among the ancient mariners: characterization of marine turtle hybrids with molecular genetic assays

Karl, S.A., B.W. Bowen, and J.C. Avise. 1995. Hybridization among the ancient mariners: characterization of marine turtle hybrids with molecular genetic assays. Journal of Heredity 86:262–268.

ANECDOTE OR BACKDROP

Within various vertebrate groups (such as birds and frogs), species separated for long periods of evolutionary time (several million years and more) nonetheless sometimes retain the anatomical and physiological capacity to produce viable hybrids. During the course of our phylogeographic surveys of marine turtles, we happened upon several interspecific hybrid animals whose parents apparently belonged to evolutionary lineages that had been separated for tens of millions of years! This article provides our genetic evidence for these seemingly outrageous phenomena.

ABSTRACT

Reports of hybridization between marine turtle species (family Cheloniidae) have been difficult to authenticate based solely on morphological evidence. Here we employ molecular genetic assays to document the sporadic, natural occurrence of viable interspecific hybrids between species representing four of the five genera of cheloniid sea turtles. Using multiple DNA markers from single-copy nuclear loci, eight suspected hybrids (based on morphology) were confirmed to be the products of matings involving the loggerhead turtle (*Caretta caretta*) × Kemp's ridley (*Lepidochelys kempii*) ($N = 1$ specimen), loggerhead turtle × hawksbill (*Eretmochelys imbricata*) ($N = 2$), loggerhead turtle X green turtle (*Chelonia mydas*) ($N = 4$), and green turtle × hawksbill ($N = 1$). Molecular markers from mitochondrial DNA permitted identification of the maternal parental species in each cross. The species involved in these hybridization events represent evolutionary lineages thought to have separated 10–75 million years ago (mya) and thus may be among the oldest vertebrate lineages capable of producing viable hybrids in nature. In some cases, human intervention with the life cycles of marine turtles (e.g., through habitat alteration, captive rearing, or attempts to establish new breeding sites) may have increased the opportunities for interspecific hybridization.

ADDENDUM

These are among the oldest ages known for the evolutionary separations between species still capable of producing viable hybrids in nature.

10

Freshwater Turtles

INTRODUCTION

Unlike marine turtles, freshwater turtles inhabit an environment that is spatially highly fractured. What is the impact of this physical structure on the population genetic architecture of turtle species that inhabit rivers and lakes? That was the basic question that motivated DeEtte Walker to join JCA's laboratory in the early 1990s and therein begin to survey freshwater turtles from across the southeastern United States. The results proved to be of interest not only in their own right but also when evaluated in comparison to emerging phylogeographic patterns also being uncovered (at that same time in JCA's lab) on freshwater fishes and terrestrial biotas in that same region of the country (see Chapters 2, 4, and 6). Several of the abstracts in this chapter summarize DeEtte's phylogeographic findings on freshwater turtles; others describe work (spearheaded by another Ph.D. student, Devon Pearse) that involved molecular assessments of genetic parentage in freshwater turtle species. This latter topic is of special interest in part because female turtles proved to be capable of storing and utilizing viable sperm (from earlier matings) across multiple years.

A comparison of mtDNA restriction sites vs. control region sequences in phylogeographic assessment of the musk turtle (*Sternotherus minor*)

Walker, D., V.J. Burke, I. Barák, and J.C. Avise. 1995. A comparison of mtDNA restriction sites vs. control region sequences in phylogeographic assessment of the musk turtle (Sternotherus minor). Molecular Ecology 4:365–373.

ANECDOTE OR BACKDROP

DeEtte Walker was a conscientious and highly competent woman who earned her Ph.D. in JCA's lab and then stayed on for several years as the laboratory's head technician. This paper marks the first in a long series of articles from DeEtte's dissertation work on phylogeographic patterns in freshwater turtles of the southeastern United States.

ABSTRACT

A total of nearly 800 base pairs of mitochondrial DNA sequence was assayed in each of 52 musk turtles (*Sternotherus minor*) collected across the species' range in the southeastern United States. About one-half of the sequence information in effect was accessed by conventional recognition site assays of the entire mtDNA molecule; the remainder came from direct sequence assays of a normally hypervariable 5′ section of the noncoding control region. The two assay methods produced essentially nonoverlapping sets of variable character states that were compared with respect to magnitudes and phylogeographic patterns of mtDNA variation. The two assay procedures yielded nearly identical outcomes with regard to: (i) total levels of species-wide mtDNA genetic variation; (ii) mean levels of within-locale variation; (iii) extremely high population genetic structure; (iv) a phylogenetically significant separation of samples from the northwestern half of the species' range versus those in the southeastern segment; and (v) considerably lower genetic variability within the northwestern clade. The micro- and macro-phylogeographic mtDNA patterns in the musk turtle are consistent with a low-dispersal natural history and with a suspected longer term biogeographic history of the species, respectively.

FIGURE 10.1   Musk turtle, *Sternotherus minor*.

Mitochondrial DNA phylogeography and subspecies issues in the monotypic freshwater turtle *Sternotherus odoratus*

Walker, D., W.S. Nelson, K.A. Buhlmann, and J.C. Avise. 1997. Mitochondrial DNA phylogeography and subspecies issues in the monotypic freshwater turtle Sternotherus odoratus. *Copeia 1997:16–21.*

ANECDOTE OR BACKDROP

This was the second in DeEtte's series of dissertation chapters. In this and in her other turtle projects, she roamed the southeastern United States laboriously trapping turtles from ponds, marshes, and swamps, in all kinds of weather.

ABSTRACT

Phylogeographic variation in mitochondrial (mt) DNA restriction sites is described for populations of the monotypic stinkpot turtle (*Sternotherus odoratus*) from the southeastern United States. Stinkpots display pronounced and hierarchical mtDNA phylogeographic differentiation, ranging from genetically shallow differences among local populations to genetically deep distinctions among regional assemblages of haplotypes. Both magnitude and general pattern of intraspecific mtDNA phylogeography in *S. odoratus* are remarkably similar to those in a related species (*Sternotherus minor*) that traditionally has been considered ditypic based on morphological and genetic appraisals. The pronounced phylogeographic population structure in *S. odoratus* effectively falsifies prior hypotheses that extensive dispersal and gene flow account for the morphometric conservatism among geographic populations. These findings also raise broader issues concerning the significance of subspecies designations in testudine systematics.

ADDENDUM

The stinkpot (S. odoratus) gets its name from the fact that the animal can release a foul musky odor from scent glands along the edge of its shell.

Phylogenetic distinctiveness of a threatened aquatic turtle (*Sternotherus depressus*)

Walker, D., G. Ortí, and J.C. Avise. 1998. Phylogenetic distinctiveness of a threatened aquatic turtle (Sternotherus depressus). **Conservation Biology** 12:639–645.

ANECDOTE OR BACKDROP

Although most of DeEtte's work involved pure (as opposed to applied) research on the phylogeographic histories of freshwater turtles, sometimes her efforts also had practical payoffs for conservation efforts. This article provides one example, involving a federally threatened species that previously had been of highly questionable taxonomic status.

ABSTRACT

The musk turtle (*Sternotherus minor*) is common throughout the southeastern United States. In 1955, a morphologically atypical form confined to the Black Warrior River System in Alabama was discovered and accorded full species status as *S. depressus*, the "flattened musk turtle." However, questions remain about the taxonomic status and evolutionary history of the flattened musk turtle because (i) the geographic distribution of *S. depressus* is nested within the range of *S. minor*; (ii) the flattened shell might be a recently evolved antipredator adaptation; and (iii) reports exist of intergrades between *S. depressus* and *S. minor*. We generated and employed sequence data from mitochondrial DNA to address the phylogenetic distinctiveness and phylogeographic position of

S. depressus relative to all other musk and mud turtles (Kinosternidae) in North America. The flattened musk turtle forms a well-supported monophyletic group separate from *S. minor*. Genetic divergence observed between *S. depressus* and geographic populations of *S. minor* is no less than that distinguishing many kinosternid congeners from one another. These molecular genetic findings bolster rationale for the taxonomic recognition of *S. depressus* and, hence, for special efforts to protect this federally threatened species.

ADDENDUM

This species remains critically endangered according to guidelines of the IUCN (International Union for Conservation of Nature).

Phylogeographic patterns in *Kinosternon subrubrum* and *K. baurii* based on mitochondrial DNA restriction analyses

Walker, D., P.E. Moler, K.A. Buhlmann, and J.C. Avise. 1998. Phylogeographic patterns in Kinosternon subrubrum *and* K. baurii *based on mitochondrial DNA restriction analyses.* **Herpetologica** *54:174–184.*

ANECDOTE OR BACKDROP

This study was another component part of DeEtte's buildup for an eventual overview of comparative phylogeographic patterns for nearly all freshwater turtles in the southeastern United States.

ABSTRACT

We used restriction assays of mitochondrial (mt) DNA to estimate phylogeographic variation in two sister taxa of mud turtles in the southeastern United States. Extensive mtDNA variation characterized *Kinosternon subrubrum* and, to a lesser degree, *K. baurii*. Each of 26 mtDNA haplotypes from the 83 assayed specimens was localized spatially. Collectively, these mtDNA haplotypes demarcated four major matrilineal assemblages, each with a well-defined regional distribution: a western group (A) in Missouri and Louisiana, a central group (B) throughout the Gulf coastal states, an eastern group (C) along the Atlantic coastal states north of Florida, and a southern group (D) in peninsular Florida. All assayed samples of *K. baurii* belonged to the mtDNA C assemblage. The two species in Florida are thus highly distinct in mtDNA genotype, but they exhibit minimal mtDNA divergence along the Atlantic coastal states. These findings raise questions concerning the evolutionary history and taxonomy of these two recognized species. mtDNA phylogeographic patterns in the *baurii/subrubrum* complex are remarkably similar to those reported previously for two other southeastern kinosternids, *Sternotherus minor* and *S. odoratus*.

FIGURE 10.2   Eastern mud turtle, *Kinosternon subrubrum*.

Phylogeographic uniformity in mitochondrial DNA of the snapping turtle (*Chelydra serpentina*)

Walker, D., P.E. Moler, K.A. Buhlmann, and J.C. Avise. 1998. Phylogeographic uniformity in mitochondrial DNA of the snapping turtle (Chelydra serpentina). *Animal* Conservation 1:55–60.

ANECDOTE OR BACKDROP

By the time this article was published, DeEtte had documented substantial phylogeographic structure in species after species of freshwater turtle that she had examined for mitochondrial DNA variation. Thus, to find a turtle species that essentially lacked mtDNA variability throughout much of its range came as a complete surprise. Although we can speculate, we still don't know why the snapping turtle alone provides such a stark exception to otherwise interpretable phylogeographic patterns in so many other freshwater turtle species.

ABSTRACT

Previous studies have revealed considerable genetic variation, geographic localization, and genealogical depth for mitochondrial DNA haplotypes within each of several species of freshwater turtles in the southeastern United States. Here we report a notable exception to such phylogeographic patterns. In control region sequences of 66 snapping turtles (*Chelydra serpentina*) collected from 10 southeastern states, a single mtDNA haplotype predominated and the two rare variants detected were nearly identical to the common genotype. This pattern of low mtDNA variation *and* a lack of appreciable geographic population structure is extremely unusual for a widely distributed animal species. For purposes of taxonomy and conservation, these findings suggest the presence of only one "evolutionarily significant unit" for *C. serpentina* in this otherwise phylogeographically rich region of the country. Possible explanations for this phylogeographic pattern in the Common Snapping Turtle are considered.

ADDENDUM

Oddly, the Alligator Snapping Turtle (Macrochelys temminckii)—a rarer and less broadly distributed relative of the Common Snapping Turtle—seems to display much higher levels of mtDNA population structure and nuclear genetic variability than does the Common Snapping Turtle.

FIGURE 10.3   Snapping turtle, *Chelydra serpentina*.

Principles of phylogeography as illustrated by freshwater and terrestrial turtles in the southeastern United States

Walker, D. and J.C. Avise. 1998. Principles of phylogeography as illustrated by freshwater and terrestrial turtles in the southeastern United States. **Annual Review of Ecology and Systematics** *29:23−58.*

ANECDOTE OR BACKDROP

This abstract is from a substantive review in which we pulled together all of DeEtte's genetic findings on numerous turtle species in the southeastern United States. These data were also compared to phylogeographic information previously compiled for freshwater fishes from the same region of the country.

ABSTRACT

Geographic patterns in mitochondrial (mt) DNA variation are compiled for 22 species of freshwater and terrestrial turtles in the southeastern United States, and the results employed to evaluate phylogeographic hypotheses and principles of genealogical concordance previously derived from similar analyses of other vertebrates in the region. The comparative molecular findings are interpreted in the context of intraspecific systematics for these taxa, the historical geology of the area, traditional nonmolecular zoogeographic information, and conservation significance. A considerable degree of phylogeographic concordance is registered with respect to: (i) the configuration of intraspecific mtDNA subdivisions across turtle species; (ii) the principle molecular partitions and traditional morphology-based taxonomic boundaries; (iii) genetic patterns in turtles versus those previously described for freshwater fishes and terrestrial vertebrates in the region; and (iv) intraspecific molecular subdivisions versus the boundaries between major zoogeographic provinces as identified by composite species' ranges in the Testudines. Findings demonstrate shared elements in the biogeographic histories of a diverse regional biota. Such phylogeographic concordances (and discordances) have ramifications for evolutionary theory as well as for the pragmatic efforts of taxonomy and conservation biology.

Genetic markers substantiate long-term storage and utilization of sperm by female painted turtles

Pearse, D.E., F.J. Janzen, and J.C. Avise. 2001. Genetic markers substantiate long-term storage and utilization of sperm by female painted turtles. **Heredity 86:378–384.**

ANECDOTE OR BACKDROP

In various animal species, mated females can store viable sperm in their reproductive tracts for periods ranging from hours to years. Furthermore, if anecdotal reports are to be believed, some females use long-stored sperm to fertilize their eggs (because females in zoos sometimes produce offspring long after their most recent known tryst with a male). Among captive vertebrates, the record duration for female sperm storage and usage—several years—probably belongs to a snake or turtle. Are such cases aberrations or do females in nature utilize long-stored sperm also? Before this article was published in 2001, nobody knew for sure.

ABSTRACT

Most studies of genetic parentage in natural populations have been limited to a single breeding season or reproductive episode and, thus, provide only a snapshot of individuals' mating behaviors. Female turtles can store viable sperm in their reproductive tracts for as long as several years, but the extent to which this capacity is utilized in nature has remained unknown. Here, we employ microsatellite markers to assess genetic paternity in successive clutches of individually marked, free-ranging female painted turtles (*Chrysemys picta*) over a 4-year period. The genetic data from 113 clutches from this natural population demonstrate that most females (80.5%) remated each year and that each female generally used a single male's sperm to fertilize all clutches laid within a year. However, sperm usage among females varied considerably, and some females apparently used sperm that had been stored for up to 3 years to fertilize some or all eggs layed in consecutive nesting seasons. Thus, remating by females is not necessary for continued offspring production from a given sire. Furthermore, 13.2 percent of all clutches examined showed evidence of multiple paternity, and the genetic paternity patterns across years suggest a "last in, first out" operation of the females' sperm storage tubules.

ADDENDUM

Long-term sperm storage by females presumably increases the opportunities for intrafemale "sperm competition" and "cryptic female choice of sperm," two prezygotic biological phenomena that have been of extraordinary scientific interest for the last two decades.

FIGURE 10.4   Painted turtle, *Chrysemys picta*.

A genetic analogue of "mark-recapture" methods for estimating population size: an approach based on molecular parentage assessments

Pearse, D.E., C.M. Eckerman, F.J. Janzen, and J.C. Avise. 2001. A genetic analogue of "mark-recapture" methods for estimating population size: an approach based on molecular parentage assessments. Molecular Ecology 10:2711–2718.

ANECDOTE OR BACKDROP

Our work on the genetics of painted turtles (see also the previous and the following abstracts) was inaugurated when JCA met Fred Janzen during a seminar trip to Iowa State University. Dr. Janzen for many years had been monitoring painted turtle populations along the Mississippi River and had accumulated volumes of phenotypic and other data on these animals. At that time, Devin Pearse was a new graduate student in JCA's laboratory, and he was assigned the task of genetically analyzing the tissue samples that Janzen so laboriously had collected. Devon devised some highly innovative projects on his own as well, as this next abstract well illustrates. In the field of wildlife management, "mark-recapture" is a standard procedure for estimating population size in species that otherwise are difficult to observe in the wild. The traditional "marks" are physical tags such as leg bands or rings that researchers fasten to captured specimens. But Mother Nature has applied her own molecular genetic tags to each and every creature. This article spearheaded by Devon shows how he took advantage of this fact to introduce a novel kind of mark-recapture analysis to population biology.

ABSTRACT

Molecular polymorphisms have been used in a variety of ways to estimate both effective and local census population sizes in nature. A related approach for estimating the current size of a breeding population, explored here for the first time, is the use of genetic "marks" reconstructed for otherwise unknown parents in paternity or maternity analyses of progeny arrays. This method provides interesting similarities and contrasts to traditional mark-recapture methods based on physical tags. To illustrate, this genetic method is applied to a population of painted turtles on the Mississippi River to estimate the number of successfully breeding males. Nongenetic mark-recapture approaches were also applied to animals trapped at this location. Results demonstrate that such genetic data on parentage can be helpful not only in estimating contemporary population sizes but also in providing additional information, not present in customary mark-recapture data, about possible extended movements of breeding individuals and the size of the pool of mates that they encounter.

Multiple paternity, sperm storage, and reproductive success of female and male painted turtles (*Chrysemys picta*) in nature

Pearse, D.E., F.J. Janzen, and J.C. Avise. 2002. Multiple paternity, sperm storage, and reproductive success of female and male painted turtles (Chrysemys picta) in nature. Behavioral Ecology and Sociobiology 51:164–171.

ANECDOTE OR BACKDROP

This is another of the Painted Turtle papers resulting from Devon Pearse's dissertation work based on our collaboration with Fred Janzen.

ABSTRACT

When females receive no direct benefits from multiple matings, concurrent multiple paternity is often explained by indirect genetic benefits to offspring. To examine such possibilities, we analyzed genetic paternity for 1,272 hatchlings, representing 227 clutches, from a nesting population of painted turtles (*Chrysemys picta*) on the Mississippi River. Goals were to quantify the incidence and distribution of concurrent multiple paternity across clutches, examine temporal patterns of sperm storage by females, and deduce the extent to which indirect benefits result from polyandrous female behaviors. Blood samples from adult males also allowed us to genetically identify the sires of surveyed clutches and to assess phenotypic variation associated with male fitness. From the genetic data, female and male reproductive success were deduced and then interpreted together with field data to evaluate possible effects of female mating behaviors and sire identity on offspring fitness. We document that more than 30% of the clutches likely were fathered by multiple males, and that presence of multiple paternity was positively correlated with clutch size. Furthermore, the data indicate that the second male to mate typically had high paternity precedence over the first.

ADDENDUM

The Painted Turtle is the most widespread and perhaps the most common turtle species native to North America. However, recent phylogeographic analyses (by other labs) have called into question the taxonomic validity of this "species."

Turtle mating systems: behavior, sperm storage, and genetic paternity

Pearse, D.E. and J.C. Avise. 2001. Turtle mating systems: behavior, sperm storage, and genetic paternity. **Journal of Heredity 92:206–211.**

ANECDOTE OR BACKDROP

This brief abstract was from a review paper that Devon Pearse wrote summarizing his dissertation research on the genetics and behavior of painted turtles.

ABSTRACT

A recent rise of interest in the mating systems of poikilothermic vertebrates has focused primarily on fishes, a few amphibians, and squamate reptiles. Turtles by contrast have received relatively little attention, yet they display a wide variety of mating behaviors and life history characteristics that make them excellent candidates for addressing several aspects of genetic parentage that should contribute to a broader understanding of animal reproductive strategies. Here we focus on genetic studies of the mating systems and reproductive patterns of turtles, specifically with respect to multiple paternity and long-term sperm storage. These phenomena highlight the importance of a temporally extended perspective on patterns of individual reproductive success.

CHAPTER

11

Other Reptiles

INTRODUCTION

Although reptiles have not been a particular focus of JCA's laboratory, his group has surveyed a modest number of reptilian species in addition to the marine and freshwater turtles already described in Chapters 9 and 10, respectively. These additional species include a potpourri of reptilian taxa ranging from the desert tortoise and desert iguana to the chuckwalla, to the estuarine diamondback terrapin, and to several species of marine sea snakes. The results again exemplify how the tools of molecular biology can be adapted to provide comparative genetic analyses of a diverse array of creatures that otherwise can be extremely difficult to study from field observations alone.

Phylogeographic patterns in mitochondrial DNA of the desert tortoise (*Xerobates agassizi*), and evolutionary relationships among the North American gopher tortoises

Lamb, T., J.C. Avise, and J.W. Gibbons. 1989. Phylogeographic patterns in mitochondrial DNA of the desert tortoise (Xerobates agassizi), *and evolutionary relationships among the North American gopher tortoises.* Evolution 43:76–87.

ANECDOTE OR BACKDROP

I have already mentioned (see Chapter 8) that Trip Lamb was an avid herpetologist even before he joined JCA's laboratory as a graduate student in the mid-1980s. Although most of his dissertation dealt with Hyla tree frogs, Trip managed to find time to conduct several other genetic analyses as well. This first abstract is from a phylogeographic study he conducted on endangered tortoises mostly native to deserts of the American southwest.

ABSTRACT

Restriction-fragment polymorphisms in mitochondrial (mt) DNA were used to evaluate population genetic structure in the desert tortoise *Xerobates agassizi* and to clarify evolutionary affinities among species of the gopher tortoise complex. Fourteen informative endonucleases were employed to assay mtDNAs from 56 *X. agassizi* representing 22 locations throughout the species' range. The mtDNA genotypes were readily partitioned into three major phylogenetic assemblages each with a striking geographic orientation. Overall, the *X. agassizi* mtDNA genotypes typify a common phylogeographic pattern wherein broad genetic uniformity of populations is interrupted by geographic features that presumably have functioned as dispersal barriers. The geologic history of the Colorado River area, which includes extensive marine incursions, may account for the marked mtDNA divergence between eastern and western *X. agassizi* assemblages. In mtDNA comparisons among the four species of the gopher tortoise complex, both UPGMA and Wagner parsimony analysis strongly support the recognition of two distinct species groups previously suggested by traditional systematic approaches. Furthermore, the mtDNA data identify the eastern *X. agassizi* assemblage as the probable inceptive lineage of *X. berlandieri*. Results from both intra- and interspecific comparisons illustrate how clues to historical geological events may be present in the geographic structure of mtDNA phylogenies.

ADDENDUM

The Desert Tortoise is listed as "vulnerable" under IUCN guidelines, thus making the taxonomy of this set of populations of special interest to conservationists.

FIGURE 11.1 Desert tortoise, *Xerobates agassizi*.

DNA fingerprints from hypervariable mitochondrial DNA genotypes

Avise, J.C., B.W. Bowen, and T. Lamb. 1989. DNA fingerprints from hypervariable mitochondrial DNA genotypes. **Molecular Biology and Evolution** *6:258–269.*

ANECDOTE OR BACKDROP

Mitochondrial DNA is notoriously polymorphic in its nucleotide sequences, but at times it can be a truly hypervariable molecule, as for example when size polymorphisms are superimposed on the sequence differences. This article describes two species in which mtDNA was so hypervariable as to qualify as providing a "DNA fingerprint" unique to each assayed specimen.

ABSTRACT

Conventional surveys of restriction-fragment polymorphisms in mitochondrial DNA of chuckwalla lizards (*Sauromalus obesus*) and menhaden fish (*Brevoortia tyrannus/patronus* complex) revealed exceptionally high levels of genetic variation, attributable to differences in mtDNA sizes as well as restriction sites. The observed probabilities that any two randomly drawn individuals differed detectably in mtDNA genotype were 0.983 and 0.998 in the two species, respectively. Thus, the variable gel profiles provided unique mtDNA "fingerprints" for most conspecific animals assayed. mtDNA fingerprints differ from nuclear DNA fingerprints in several empirical respects and should find special application in the genetic assessment of maternity.

ADDENDUM

The findings of this study are not to be confused with conventional "DNA fingerprints" that come from hypervariable segments of the freely recombining nuclear genome.

Phylogeographic histories of representative herpetofauna of the southwestern U.S.: mitochondrial DNA variation in the desert iguana (*Dipsosaurus dorsalis*) and the chuckwalla (*Sauromalus obesus*)

Lamb, T., T.R. Jones, and J.C. Avise. 1992. Phylogeographic histories of representative herpetofauna of the southwestern U.S.: mitochondrial DNA variation in the desert iguana (Dipsosaurus dorsalis) *and the chuckwalla* (Sauromalus obesus). **Journal of Evolutionary Biology** *5:465–480.*

ANECDOTE OR BACKDROP

By the early 1990s, phylogeographic surveys had been conducted by JCA's laboratory on numerous freshwater, terrestrial, and maritime species in the southeastern United States. Furthermore, several concordant phylogeographic patterns had emerged, suggesting shared historical responses to earlier climatic and physiographic forces. Might concordant phylogeographic patterns be present in other regional faunas as well? We already had demonstrated a sharp genetic discontinuity in desert tortoises of the American southwest, so a logical follow-up question was whether juxtaposed discontinuities might be apparent in other desert dwelling species as well. This study by Trip Lamb was designed to offer an initial test of this possibility.

ABSTRACT

To determine whether genetic variation in representative reptiles of the southwestern United States may have been similarly molded by the geologic history of the lower Colorado River, we examined restriction site polymorphisms in the mitochondrial (mt) DNA of desert iguanas (*Dipsosaurus dorsalis*) and chuckwallas (*Sauromalus obesus*). Observed phylogeographic structure in these lizards was compared to that reported for the desert tortoise (*Xerobates agassizi*), whose mtDNA phylogeny demonstrates a striking genetic break at the Colorado River. Both the desert iguana and chuckwalla exhibit extensive mtDNA polymorphism, with respective genotypic diversities $G = 0.963$ and 0.983, close to the maximum possible value of 1.0. Individual mtDNA clones, as well as clonal assemblages defined by specific levels of genetic divergence, showed pronounced geographic localization. Nonetheless, for each species the distributions of particular clones and most major clonal groupings encompass both sides of the Colorado River valley, and hence are clearly incongruent with the phylogeographic pattern of the desert tortoise. Overall, available molecular evidence provides no indication that the intraspecific phylogenies of the southwestern US herpetofauna have been concordantly shaped by a singular vicariant factor of overriding significance.

ADDENDUM

In these preliminary surveys we were disappointed not to find more phylogeographic concordance across taxa in the American Southwest. Nevertheless, many more species from this region should be genetically studied and compiled in a comparative vein. The chuckwallas and desert iguanas were included in this study simply because of their availability, and because they have suitable species' ranges across deserts of the American southwest. The generic name of the chuckwalla (Sauromalus) comes from the ancient Greek words "sauros" meaning lizard and "omalus" meaning flat. Each member of this species can indeed flatten its otherwise obese body to squeeze into small crevices in rocks.

FIGURE 11.2   Chuckwalla, *Sauromalus obesus*.

Molecular and population genetic aspects of mitochondrial DNA variability in the diamondback terrapin, *Malaclemys terrapin*

Lamb, T. and J.C. Avise. 1992. Molecular and population genetic aspects of mitochondrial DNA variability in the diamondback terrapin, Malaclemys terrapin. *Journal of Heredity 83:262–269.*

ANECDOTE OR BACKDROP

A major challenge in this study was simply collecting these beautiful animals to begin with. This was eventually accomplished by a variety of tedious methods including intensive seining, trapping, gillnetting, dipnetting, and purchasing specimens from local fishers. The project proved worthy of the effort, as a shallow but sharp genetic separation was detected between populations inhabiting the Atlantic versus Gulf coasts of the southeastern United States.

ABSTRACT

Diamondback terrapins (*Malaclemys terrapin*) occupy brackish waters along North America's Atlantic and Gulf coasts. Despite a nearly continuous distribution, terrapin populations exhibit extensive geographic variation, with seven subspecies recognized. To assess population genetic structure in *Malaclemys*, we used 18 restriction enzymes to assay mitochondrial (mt) DNA genotypes in 53 terrapins collected from Massachusetts to western Louisiana. mtDNA size polymorphism and heteroplasmy were observed, attributable to variation in copy number of a 75-bp tandem repeat. In terms of restriction sites, mtDNA genotypic diversity ($G = 0.582$) and divergence levels ($p < 0.004$) were exceptionally low. Only one restriction site polymorphism appeared geographically informative, cleanly distinguishing populations north versus south of Florida's Cape Canaveral region. Nonetheless, the probable zoogeographic significance of this single site change is underscored by its (i) perfect concordance with the distribution of a key morphological character and (ii) striking agreement with the phylogeographic patterns observed for mtDNA profiles of several other coastal marine species. The possible isolation of Atlantic and Gulf terrapin populations during late-Pleistocene glacial maxima conceivably accounts for the observed patterns of mtDNA (and morphological) variation.

ADDENDUM

The Diamondback Turtle is a rare example of a turtle species that prefers brackish waters.

FIGURE 11.3   Diamondback Terrapin, *Malaclemys terrapin*.

Genetic monandry in six viviparous species of true sea snakes

Lukoschek, V. and J.C. Avise. 2011. Genetic monandry in six viviparous species of true sea snakes. **Journal of Heredity** *102:347–351.*

ANECDOTE OR BACKDROP

In the late 2000s, a postdoc from Australia joined JCA's lab. Vimoksalehi Lukoschek was already an expert on sea snakes (in Elapidae), by far the most species-rich of all marine reptiles, and JCA had visions of starting a multifaceted project on these venomous and viviparous animals (in a research program that might be quite analogous to that already conducted on marine turtles [see Chapter 9*]). Alas, for a variety of logistical and other reasons, such a multidimensional genetic and phylogeographic program on sea snakes has yet to materialize. Nevertheless, Vimoksalehi did conduct and publish several preliminary studies, the first two of which are of summarized in the following pair of abstracts.*

ABSTRACT

Using a suite of 10 highly variable microsatellite loci, we conducted genetic paternity analyses for 76 embryos in the broods of 12 pregnant females representing 6 viviparous species of true sea snakes (the *Hydrophis* clade) in the family Elapidae. To our surprise, we uncovered no evidence for multiple paternity within any of the clutches despite the fact that the genetic markers showed high intraspecific heterozygosities and as many as 20 conspecific alleles per locus. This outcome stands in sharp contrast to the rather high (but also variable) frequency of multiple paternity previously reported in many other reptilian species. However, because our study appears to be the first assessment of genetic parentage for any sea snake species (and indeed for any member of the elapid clade), we can only speculate as to whether this apparent monandry by females is a broader phylogenetic characteristic of elapid snakes or whether it might relate somehow to the sea snakes' peculiar lifestyle that uniquely combines viviparity with a marine existence.

ADDENDUM

Although sea snakes are highly venomous, they are generally nonaggressive to humans and not particularly dangerous in the wild (unless mishandled). All sea snakes generally resemble eels. But unlike eels and other fish, sea snakes do not have gills, and hence must come to the surface regularly to breathe. Sea snakes are closely related to some terrestrial venomous snakes in Australia.

FIGURE 11.4   Elegant sea snake, *Hydrophis elegans*.

Development of eleven polymorphic microsatellite loci for the sea snake *Emydocephalus annulatus* (Elapidae: Hydrophiinae) and cross-species amplification for seven species in the sister genus *Aipysurus*

Lukoschek, V. and J.C. Avise. 2012. Development of eleven polymorphic microsatellite loci for the sea snake Emydocephalus annulatus *(Elapidae: Hydrophiinae) and cross-species amplification for seven species in the sister genus* Aipysurus. *Conservation Genetics Resources 4:11−14.*

ABSTRACT

We developed eleven microsatellite loci for the turtle headed sea snake, *Emydocephalus annulatus*, from partial genomic DNA libraries using a repeat enrichment protocol. Nine loci had high numbers of alleles (11−32) per locus, while the other two loci had six alleles each. All 11 loci amplified successfully and were polymorphic in 6 of 7 sea snake species from the sister genus *Aipysurus*, while 10 loci amplified successfully for the seventh species. Based on these highly successful cross-amplifications, we expect these 11 loci to be useful markers for evaluating population genetic structure, gene flow, relatedness, and effective population sizes for all *Aipysurus* group sea snakes.

ADDENDUM

Several dozen extant species of true sea snakes exist, comprising the distinct Hydrophis and Aipysurus evolutionary lineages.

12

Birds

INTRODUCTION

JCA has had a lifelong love affair with birds. He began birdwatching at the tender age of six and has continued this hobby throughout his life, birding and photographing birds from around the world even while on seminar trips. Birds have also been an important part of JCA's scientific career. He has taught Ornithology for more than 40 years at the Universities of Georgia and California. He and his students have researched birds in a variety of contexts ranging from appraisals of population genetic structure and gene flow to phylogenetic relationships among close and distant species. He is an elected fellow of the American Ornithologists' Union and he has received a William Brewster Memorial Award from that organization for career contributions to avian biology.

The abstracts gathered in this chapter reflect JCA's diverse research on the class Aves. They begin with early allozyme surveys and continue well into the phylogeographic era with appraisals of mtDNA. An important event occurred early in JCA's career when he became acquainted with the Tall Timbers Research Station near Tallahassee, Florida. This ecological field station is located immediately adjacent to an 1100-foot-tall television tower that routinely knocked down (and thereby inadvertently killed) many hundreds of birds during their annual spring and fall migrations. Researchers at the field station collected these avian corpses and stored them in a freezer for the preparation of museum skins and for other scientific uses. What had been a tragedy for the avian migrants became a bonanza for JCA's early research, because frozen tissues from many otherwise difficult-to-collect bird species were made available for genetic assays. Thus, much of the early avian research in JCA's lab involved genetic analyses of birds that were stored in the Tall Timber's morgue freezer.

Evolutionary genetics of birds. I. Relationships among North American thrushes and allies

Avise, J.C., J.C. Patton, and C.F. Aquadro. 1980. Evolutionary genetics of birds. I. Relationships among North American thrushes and allies. The Auk 97:135–147.

ANECDOTE OR BACKDROP

This was the first in a series of protein-electrophoretic surveys of various avian taxa from which multiple individuals sadly had been killed during migration by the Tall Timbers television tower. All of the papers in this series were among the earliest such multi-locus allozyme surveys for any bird species.

ABSTRACT

We have employed phenetic and cladistic approaches to analyze frequencies of electromorphs encoded by 25–27 loci in several species of North American thrushes (family Muscicapidae) and their relatives. In broad perspective both methods of data analysis yield similar summaries of probable relationships among these species. All four examined species of *Catharus* are nearly identical in electromorph composition. *Hylocichla mustelina* is phenetically and cladistically allied to members of *Catharus*. Both *Turdus migratorius* and *Sialia sialis* lie outside the *Hylocichla-Catharus* clade and are phenetically quite distinct from it and from each other. Despite its current placement in a distinct family, Mimidae, *Dumatella carolinensis* is cladistically and phenetically more closely allied to the Turdinae than is *Regulus calendula*, in some classifications a current member of the Muscicapidae, subfamily Sylviinae. A major point of ambiguity in our data concerns the relative affinities of *Turdus migratorius* and *Sialia sialis* to the *Hylocichla-Catharus* clade. Phenetically, both are roughly equidistant to that clade, and cladistically our data can yield alternative interpretations. The levels of genetic similarity between muscicapid species at various stages of taxonomic divergence are comparable to previous estimates for other Passeriformes but are far higher than typical estimates for other vertebrate and invertebrate taxa of equal rank.

ADDENDUM

Nocturnal migrants are most vulnerable to the lighted TV tower, and Catharus thrushes are nocturnal migrants; hence their availability for this early phylogenetic survey. For example, on the night of May 2, 1964, more than 100 Gray-cheeked Thrushes (Catharus minimus) were found dead at the base of the Tall Timbers television tower; and >1000 Veerys (Catharus fuscescens) have been killed at this same location over the years.

FIGURE 12.1 Swainson's
Thrush, *Catharus ustulatus*.

Evolutionary genetics of birds. II. Conservative protein evolution in North American sparrows and relatives

Avise, J.C., J.C. Patton, and C.F. Aquadro. 1980. Evolutionary genetics of birds. II. Conservative protein evolution in North American sparrows and relatives. Systematic Zoology 29:323–334.

ANECDOTE OR BACKDROP

This was the second in that early series of avian allozymic analyses.

ABSTRACT

Differentiation at 20−21 protein-coding genes was examined by conventional techniques of starch-gel electrophoresis among twelve species and seven genera of North American sparrows and relatives: Emberizidae, subfamily Emberizinae. One species representing Fringillidae was also included. Data were summarized in a distance matrix that was subsequently used to infer phylogenetic trees by a variety of methods. Results were generally consistent with current classification. Two salient results were unanticipated: (1) the relatively close genetic similarity of *Pipilo* to group I Emberizinae; and (2) the relatively large genetic distance of *Calcarius* from other Emberizinae. A search of the litrature revealed that the distribution of a behavioral characteristic—bilateral scratching—had led to a prediction of phylogenetic relationships for these genera that proved to be fully consistent with the protein information. This result is significant because it lends support to earlier proposals that some behavioral traits can be extremely valuable as phylogenetic markers. Levels of protein divergence in birds are compared to previous estimates for other vertebrate taxa. At corresponding levels of the taxonomic hierarchy, birds consistently exhibit far smaller genetic distances than do many fishes and other vertebrates.

ADDENDUM

This was one of the earliest studies documenting that behavioral traits could be used in conjunction with allozyme markers to suggest phylogenetic relationships among related species. The sparrows assayed in this project again were tower-kill birds from the Tall Timbers site. During the 20 years in which the TV tower grounds were surveyed for avian corpses, an average of 1600 birds were killed annually at just this single location. Indeed, the kills occurred virtually every night from mid-August to mid-November during the autumnal migration, and then again during the spring. Over the years, a total of more than 42,000 tower-kill birds were counted by research staff at the Tall Timbers Research Station. Sometimes particular bird species were devastated at specific points in time, especially when the weather conditions were conducive to migration. For example, on the night of October 9, 1955, about 200 Palm Warblers (plus an estimated 5,000 individuals of various other species) died at the TV tower, making this the largest recorded avian kill in the tower's sad history. When one considers the total number of communication towers (and tall buildings) in the U.S. alone, the total numbers of birds killed by collisions each year must be truly astronomical.

FIGURE 12.2 Song Sparrow, *Melospiza melodia*.

Evolutionary genetics of birds. V. Genetic distances within Mimidae (mimic thrushes) and Vireonidae (vireos)

Avise, J.C., C.F. Aquadro, and J.C. Patton. 1982. Evolutionary genetics of birds. V. Genetic distances within Mimidae (mimic thrushes) and Vireonidae (vireos). **Biochemical Genetics 20:95–104.**

ANECDOTE OR BACKDROP

And this was the fifth in that series of protein-electrophoretic papers, from which an emerging generalization was that avian taxa at a given level of taxonomic recognition often exhibit much smaller genetic distances at protein-coding loci than do many of their non-avian taxonomic counterparts.

ABSTRACT

Genetic distances (D's) between five species within each of the families Mimidae and Vireonidae were estimated from frequencies of protein electromorphs at 23 loci. For three mimid species in the genus *Toxostoma*, mean D equals 0.084 (range 0.069–0.104); and among three mimid genera, mean D equals 0.223 (0.167–0.278). These distances typify values previously reported in other birds at comparable levels of taxonomic recognition. In sharp contrast, the mean genetic distance among five congeneric species of Vireonidae is far higher, mean $D = 0.360$ (0.027–0.578). One possible explanation for these results is that *Vireo* species are considerably older, on the average, than are species of *Toxostoma* or than are members of several other avian genera assayed to date. Conventional thought about the origin and relative age of the Vireonidae appears compatible with this explanation. Although genetic distances in the Vireonidae are large by "avian standards," they remain modest or even small in comparison with distances between many nonavian vertebrate congeners. Results for the Mimidae and the Vireonidae are directly contrasted with genetic distances in well-known genera of Amphibia and Reptilia.

FIGURE 12.3   Northern Mockingbird, *Mimus polyglottis*.

Evolutionary genetics of birds. VI. A reexamination of protein divergence using varied electrophoretic conditions

Aquadro, C.F. and J.C. Avise. 1982. Evolutionary genetics of birds. VI. A reexamination of protein divergence using varied electrophoretic conditions. Evolution 36:1003–1019.

ANECDOTE OR BACKDROP

One possible explanation for the "conservative" pattern of protein differentiation in birds is that the protein-electrophoretic methods employed were somehow especially insensitive to real genetic differences that might exist among avian taxa. This study was designed to critically test that proposition. It was conducted by an industrious young graduate student—Chip Aquadro—who had recently joined JCA's laboratory. Chip later went on to have a highly successful scientific career in his own right.

ABSTRACT

A general pattern of conservative protein divergence among avian taxa has recently been reported in several multi-locus electrophoretic surveys. One hypothesis to account for this conservative pattern is that the particular "one-pass" electrophoretic conditions employed were, for some reason, particularly insensitive to real differences in avian proteins. Here we evaluate this possibility by systematically varying a number of electrophoretic conditions (pH, ion type, gel concentration) to assay enzyme products of six loci in North American thrushes and relatives. A total of 22 electromorphs were resolved under standard conditions, and eight additional variants were revealed by the use of 7–11 other conditions. None of the new variation occurred within members of the genus *Catharus*, or between the related gerera *Catharus* and *Hylocichla*. No single condition resolved all variants, and there was no consistent relationship between discriminatory power and buffer pH. The amount of new variation differed across loci. For two proteins previously reported to be particularly conservative in birds (the supernatant and mitochondrial forms of malate dehydrogenase), our varied-condition approach was extended to assay representatives of 22 taxonomic families in 10 avian orders. New variation was uncovered, but the total number of electromorphs remained low considering the taxonomic diversity encompassed. Overall, our results confirm and strengthen the observation of a conservative pattern of protein divergence in birds.

ADDENDUM

The technical era of "varied protein-electrophoretic conditions" passed quickly and had no major or lasting impact on the field of evolutionary genetics (which simply moved on to other more pressing issues). For example, by the mid-1980s, mitochondrial assays already had begun to supplant allozymes (protein electrophoresis) as a favored tool for many applications in molecular ecology; and microsatellite assays were soon to replace allozymes as favored polymorphic molecular markers from the nuclear genome.

An empirical evaluation of qualitative Hennigian analyses of protein electrophoretic data

Patton, J.C. and J.C. Avise. 1983. An empirical evaluation of qualitative Hennigian analyses of protein electrophoretic data. Journal of Molecular Evolution *19:244–254.*

ANECDOTE OR BACKDROP

At the same time that our early allozyme surveys of birds were being conducted, a broader revolution was taking place in systematics. I'm referring to the early rise of Hennigian cladism, which was transforming the practice of phylogenetic reconstruction. So as not to be left behind by this conceptual transformation, another young graduate student in the lab—John Patton—decided to reanalyze several genetic data sets from JCA's research program, this time using qualitative Hennigian methodologies. His findings and their interpretations are described in a paper that had the following abstract.

ABSTRACT

In an empirical evaluation of a qualitative approach to construction of phylogenetic trees from protein-electrophoretic data, we have employed Hennigian cladistic principles to generate molecular trees for waterfowl, rodents, bats, and other phylads. This procedure of tree construction is described in detail. Branching structures of molecular trees produced by three different algorithms were compared against those of "model" classifications previously proposed by other systmatists. In each case, the qualitative cladistic trees provided fits to model phylogenies that were strong and as good or better than those resulting from phenetic-clustering or distance-Wagner trees based on manipulation of quantitative values in matrices of genetic distance. The qualitative Hennigian approach has several pragmatic (as well as theoretical) advantages for analyzing routine sets of electrophoretic data: (1) the analyses are simple and can be performd by hand; (2) they provide the researcher with a strong "feel" for the data; (3) additional data (from new loci or species) can readily be added to the tree without the need to recalculate distance matrices; and (4) the qualitative output of the analyses explicitly defines character states along all branches of the tree, and hence affords a high degree of testability. However, these advantages are counterbalanced by a number of serious disadvantages that will likely limit the general applicability of this qualitative approach. These drawbacks are also discussed in detail.

ADDENDUM

The heated debate between the pheneticists and the cladists of the 1970s has by now faded to a distant (and rather sour) memory.

Systematic relationships among waterfowl (Anatidae) inferred from restriction endonuclease analysis of mitochondrial DNA

Kessler, L.G. and J.C. Avise. 1984. Systematic relationships among waterfowl (Anatidae) inferred from restriction endonuclease analysis of mitochondrial DNA. Systematic Zoology 33:370–380.

ANECDOTE OR BACKDROP

Lou Kessler was another of JCA's early graduate students who jumped on the avian bandwagon in those early years. His dissertation work focused mostly on the phylogenetic relationships of waterfowl, especially as assessed by mtDNA. Lou was one of relatively few of JCA's students who did not remain in basic science after graduation. Instead, he went to work as a genetic consultant for a patent law firm in Washington, D.C. I suspect that Lou may have done better financially than did his mentor and many of his compatriots who remained in academia!

ABSTRACT

To evaluate the potential of mitochondrial (mt) DNA analysis for avian systematics, we have assayed mtDNA differences among 13 species of waterfowl in the genera *Anas* and *Aythya* (Anseriformes: Anatidae). Purified mtDNA was digested with each of 15 different type II restriction endonucleases that cleave at five- or six-base recognition sequences. Side-by-side comparisons of digestion profiles permitted the estimation of levels of fragment homology and nucleotide sequence divergence (p). Among nine *Anas* species, mean sequence divergence was $p = 0.062$ (range 0.004–0.088); among four *Aythya* species, mean $p = 0.034$ (0.025–0.043); between selected species belonging to separate genera, mean $p = 0.109$. Phylogenetic trees and dendrograms were constructed from qualitative and quantitative data bases by a variety of procedures including undirected parsimony (Penny and Wagner algorithms), undirected compatibility (Estabrook algorithms), and phenetic clustering. These trees were highly concordant with one another, and with traditional phylogenies derived from independent sources of information. Previously published evidence from mammals has suggested a "saturation effect" on level of mtDNA differentiation: for $p < 0.15$–0.20, mtDNA distances are reportedly linearly related to time since common ancestry, but for large p values the relationship becomes curvilinear as differentiation approaches an observed plateau at approximately $p = 0.30$. Our estimates of mtDNA sequence divergence among congeneric waterfowl fall in a broad range well within the expected linear portion of the curve. This observation, coupled with the general concordance of mtDNA-generated trees with those derived from independent information, demonstrates that the restriction fragment approach to mtDNA analysis should provide an important new molecular technique for studying evolutionary relationships among lower taxonomic levels in Aves.

ADDENDUM

The phylogenetic utility of mtDNA sequences is now taken for granted by nearly all systematists.

FIGURE 12.4   Northern Pintail, *Anas acuta*.

A comparative description of mitochondrial DNA differentiation in selected avian and other vertebrate genera

Kessler, L.G. and J.C. Avise. 1985. A comparative description of mitochondrial DNA differentiation in selected avian and other vertebrate genera. **Molecular Biology and Evolution 2:109−125.**

ANECDOTE OR BACKDROP

This abstract was from another paper that formed a part of Lou Kessler's dissertation. It showed that the relatively conservative pattern of evolution in avian proteins extended to mtDNA sequences as well.

ABSTRACT

Levels of mitochondrial (mt) DNA sequence divergence between species within each of several avian (*Anas, Aythya, Dendroica, Melospiza,* and *Zonotrichia*) and nonavian (*Lepomis* and *Hyla*) vertebrate genera were compared. An analysis of digestion profiles generated by 13−18 restriction endonucleases indicates little overlap in magnitude of mtDNA divergence for the avian versus nonavian taxa examined. In 55 interspecific comparisons among the avian congeners, the fraction of identical fragment lengths (F) ranged from 0.26 to 0.96 (mean $F = 0.46$), and, given certain assumptions, these translate into estimates of nucleotide sequence divergence ranging from 0.007 to 0.088; in 46 comparisons among the fish and amphibian congeners, F values ranged from 0.00 to 0.36 (mean $F = 0.09$), yielding estimates of $p > 0.070$. The small mtDNA distances among avian congeners are associated with protein-electrophoretic distances (D values) less than about 0.20, while the mtDNA distances among assayed fish and amphibian congeners are associated with D values usually > 0.40. Since the conservative pattern of protein differentiation previously reported for many avian versus nonavian taxa now appears to be paralleled by a conservative pattern of mtDNA divergence, it seems increasingly likely that many avian species have shared more recent common ancestors than have their nonavian taxonomic counterparts. However, estimates of avian divergence times derived from mtDNA-calibrated and protein-calibrated clocks cannot readily be reconciled with some published dates based on limited fossil remains. If the earlier paleontological interpretations are valid, then protein and mtDNA evolution must be somewhat decelerated in birds. The empirical and conceptual issues raised by these findings are highly analogous to those in the longstanding debate about rates of molecular evolution and separation times of ancestral hominids from African apes.

Evolutionary genetics of birds. IV. Rates of protein divergence in waterfowl (Anatidae)

Patton, J.C. and J.C. Avise. 1986. Evolutionary genetics of birds. IV. Rates of protein divergence in waterfowl (Anatidae). **Genetica 68:129–143.**

ANECDOTE OR BACKDROP

This abstract was from another paper that formed a part of John Patton's dissertation.

ABSTRACT

An electrophoretic comparison of proteins in 26 species of waterfowl (Anatidae), representing two major subfamilies and six subfamilial tribes, led to the following major conclusions: (1) the genetic data, analyzed phenetically and cladistically, generally support traditional concepts of evolutionary relationships, although some areas of disagreement are apparent; (2) species and genera within Anatidae exhibit smaller genetic distances at protein-coding loci than do most non-avian vertebrates of equivalent taxonomic rank; (3) the conservative pattern of protein differentiation in Anatidae parallels patterns previously reported in passeriform birds. If previous taxonomic assignments and ages of anatid fossils are reliable, it would appear that the conservative levels of protein divergence among living species may not be due to recent age of the family, but rather to a several-fold deceleration in rate of protein evolution relative to nonavian vertebrates. Since it now appears quite possible that homologous proteins can evolve at different rates in different phylads, molecular-based conclusions about absolute divergence times for species with a poor fossil record should remain appropriately reserved. However, the recognition and study of the phenomenon of rate heterogeneity of protein evolution across phylads may eventually enhance our understanding of both molecular and organismal biology.

Malate dehydrogenase isozymes provide a phylogenetic marker for the Piciformes (woodpeckers and allies)

Avise, J.C. and C.F. Aquadro. 1987. Malate dehydrogenase isozymes provide a phylogenetic marker for the Piciformes (woodpeckers and allies). **The Auk 104:324–328.**

ANECDOTE OR BACKDROP

During the course of our protein-electrophoretic investigations of birds, JCA stumbled upon circumstantial evidence for a gene duplication for a central metabolic enzyme (malate dehydrogenase) in woodpeckers. This led to the following study in which we surveyed many woodpecker allies for presence versus absence of the unusual zymogram (gel) pattern, the intent being to document the taxonomic distribution of this duplication and thereby trace its evolutionary origins. The work entailed getting tissue samples from South American toucans and barbets, among many other exotic avian species. Despite repeated efforts we were unable to obtain collecting permits for these taxa, so instead we finally coordinated with the Natural History Museum of Louisiana State University to obtain the necessary samples (because JCA knew that these museum personnel had an ongoing research project on the South American avifauna).

ABSTRACT

Here we characterize an unusual zymogram (protein-electrophoretic) pattern for malate dehydrogenase (MDH) in the Piciformes (woodpeckers and their allies). Unlike in all other surveyed birds, the supernatant form of MDH shows three distinct bands in specimens representing 23 species of woodpeckers (Picidae), honeyguides (Indicatoridae), barbets (Capitonidae), and toucans (Rhamphastidae). This unique zymogram pattern, perhaps the result of a gene duplication, provides an informative phylogenetic marker for the Piciformes, apparently having arisen during evolution after the taxonomic families listed above split off from puffbirds (Bucconidae), jacamars (Galbulidae), and other more distantly related avian taxa.

ADDENDUM

Despite their odd appearance and exceptionally large bills, barbets and toucans were long suspected to be evolutionary allies of woodpeckers, based on detailed considerations of various other morphological features. This study provided strong molecular evidence for these suspected phylogenetic affiliations, by showing that barbets, toucans, and woodpeckers (unlike nearly all other surveyed birds) show similar evidence for a particular gene duplication, which therefore provides a compelling synapomorphy (shared-derived trait) for this particular avian clade.

FIGURE 12.5 Toucan (top branch) and two Barbet species.

Phylogeographic population structure of red-winged blackbirds assessed by mitochondrial DNA

Ball, R.M., Jr., S. Freeman, F.C. James, E. Bermingham, and J.C. Avise. 1988. Phylogeographic population structure of red-winged blackbirds assessed by mitochondrial DNA. **Proceedings of the National Academy of Sciences USA** *85:1558—1562.*

ANECDOTE OR BACKDROP

The Red-winged Blackbird is arguably the most abundant and widespread bird species native to North America. This paper describes our continent-wide phylogeography survey of this taxon, samples of which were obtained mostly by a letter-writing campaign to many of JCA's scientific colleagues around the country who could supply such tissues from their own research on this well-known animal. The broader intellectual rationale of our genetic survey was to reconcile two opposing perspectives on avian population structure. On the one hand, birds have a tremendous dispersal potential due to flight and a proclivity to migrate; but, on the other hand, individuals in many species show strong fidelity to their natal sites or regions. How do these opposing behavioral propensities play out in terms of the realized phylogeographic population structures of avian taxa vis-à-vis their terrestrial counterparts such as many small mammals? This paper was among the first to address such questions from the powerful genealogical perspective that mitochondrial DNA provides.

ABSTRACT

A continent-wide survey of restriction-site variation in mitochondrial (mt) DNA of the Red-winged Blackbird (*Agelaius phoeniceus*) was conducted to assess the magnitude of phylogeographic population structure in an avian species. A total of 34 mtDNA genotypes were observed among the 127 specimens assayed by 18 restriction endonucleases. Nonetheless, population differentiation was minor, as indicated by: (1) small genetic distances in terms of base substitutions per nucleotide between mtDNA genotypes (maximum $p = 0.008$); and by (2) the widespread geographic distributions of particular mtDNA clones and phylogenetic arrays of clones. Extensive morphological differentiation among redwing populations apparently has occurred in the context of relatively little phylogenetic separation. A comparison between mtDNA data sets for redwings and deermice (*Peromyscus maniculatus*) also sampled from across North America shows that intraspecific population structures of these two species differ dramatically. The lower phylogeographic differentiation in redwings is probably due to historically higher levels of gene flow.

ADDENDUM

The paucity of mtDNA phylogeographic structure in Red-winged Blackbirds has proved to be somewhat of an anomaly, even by the conservative avian standards. Several other broadly distributed avian species have been shown (in other labs) to be sharply structured phylogeographically, despite birds' capacity for flight.

FIGURE 12.6 Red-winged Blackbird, *Agelaius phoeniceus*.

Molecular genetic divergence between avian sibling species: King and Clapper Rails, Long-billed and Short-billed Dowitchers, Boat-tailed and Great-tailed Grackles, and Tufted and Black-crested Titmice

Avise, J.C. and R.M. Zink. 1988. Molecular genetic divergence between avian sibling species: King and Clapper Rails, Long-billed and Short-billed Dowitchers, Boat-tailed and Great-tailed Grackles, and Tufted and Black-crested Titmice. The Auk 105:516–528.

ANECDOTE OR BACKDROP

Bob Zink (who was then a faculty member at Louisiana State University) is a close personal friend of JCA, and likewise is an avid outdoorsman as well as professional ornithologist. In this article, the two teamed up to genetically compares several pairs of avian "sibling species" (closest living relatives). Members of each pair are morphologically nearly identical, yet they invariably proved to be readily distinguishable by the molecular markers employed, which included both allozymes and mitochondrial DNA.

ABSTRACT

Surveys of electrophoretic variation in proteins, and restriction site variation in mitochondrial (mt) DNA, were conducted to assess the resolving power of these molecular genetic techniques to distinguish four pairs of avian sibling taxa. Samples of rails (*Rallus elegans* and *R. longirostris*), dowitchers (*Limnodromus scolopaceus* and *L. griseus*), grackles (*Quiscalus major* and *Q. mexicanus*), and titmice (*Parus bicolor bicolor* and *P.b. atricristatus*) were assayed for allozymes encoded by 34–37 nuclear loci, and for an average of 77 mtDNA restriction sites per individual by 19 endonucleases. MtDNA's of the two rail species showed large-scale size polymorphism and individual heteroplasmy, the first such findings of these molecular features in an avian species. Genetic distances based on allozyme comparisons were small for all assayed taxa (Nei's $D < 0.063$). The mtDNA assays offered consistently greater resolving power, providing at least five restriction site differences for samples of any taxon pair. The Long-billed and Short-billed dowitchers were especially divergent, differing by at least 24 assayed mtDNA restriction sites and an estimated nucleotide sequence divergence of $p = 0.082$. We compared these results to previous reports of genetic distances within and among closely related bird species. The mtDNA divergence among dowitchers is near the high end of the scale of such estimates for avian congeners. The mtDNA distances between the pairs of rails ($p = 0.006$), titmice ($p = 0.004$), and grackles ($p = 0.016$) were typical for extremely closely related species, and overlap maximum values reported for some avian conspecifics.

ADDENDUM

Results of this study have since proved to be quite generalizable: Typically, even avian species that are extremely similar in phenotype can be readily distinguished in molecular genetic assays.

FIGURE 12.7   Clapper Rail, *Rallus longirostris.*

Molecular genetic relationships of the extinct Dusky Seaside Sparrow

Avise, J.C. and W.S. Nelson. 1989. Molecular genetic relationships of the extinct Dusky Seaside Sparrow. Science 243:646–648.

ANECDOTE OR BACKDROP

By the late 1970s, the endangered Dusky Seaside Sparrow had become an icon for the growing conservation movement in North America. Populations of this small coastal bird were in sharp decline: an intensive field survey in 1980 found only six remaining birds in the wild. These birds were brought into captivity (by wildlife officials) in what proved to be an unsuccessful attempt to establish a captive breeding population. This paper presents our evolutionary genetic discoveries about the seaside sparrow taxonomic complex. It was coauthored by Bill Nelson, a long-term technician in JCA's lab. The molecular findings caught almost everyone by complete surprise.

ABSTRACT

Mitochondrial DNA from the newly extinct dusky seaside sparrow (*Ammodramus maritimus nigrescens*) was compared in terms of nucleotide sequence divergence to mtDNAs from extant populations of seaside sparrows. Analyses of restriction sites revealed a close phylogenetic affinity of *A. m. nigrescens* to other sparrow populations along the Atlantic coast of the United States but considerable genetic distance from Gulf coast birds. Concerns and applied management strategies for the seaside sparrow have been based on a morphological taxonomy that does not adequately reflect evolutionary relationships within the complex.

ADDENDUM

The Dusky Seaside Sparrow was originally characterized and recognized as a separate taxonomic entity in 1873, based on its relatively dark plumage coloration and distinctive song. Its range was apparently confined to marshes along Florida's Atlantic coast, on Merritt Island and the upper Saint Johns River. This population officially went extinct in 1987, when the last known specimen of this race died in captivity. This now-extinct subspecies is survived by many other Seaside Sparrow populations along the Atlantic and Gulf coasts of the eastern United States. Our finding of an extremely close genetic relationship between the Dusky Seaside Sparrow and other Atlantic coast populations generated a firestorm of controversy about the validity of subspecies status for this population, and more generally about the broader relevance of genetic data for taxonomic entities granted special protection under the United States Endangered Species Act.

FIGURE 12.8   Seaside Sparrow, *Ammodramus maritimus.*

A role for molecular genetics in the recognition and conservation of endangered species

Avise, J.C. 1989. A role for molecular genetics in the recognition and conservation of endangered species. **Trends in Ecology and Evolution 4:279–281.**

ANECDOTE OR BACKDROP

Before the time when this review paper was published, wildlife officials and conservation organizations seldom gave more than lipservice to genetic concerns for threatened and endangered taxa. In this paper, JCA made a case—based on growing empirical experience—that molecular markers had much to add to conservation science. Today this sentiment is so widely accepted as to be taken nearly for granted.

ABSTRACT

Taxonomies based on morphological traits alone sometimes provide inadequate or misleading guides to phylogenetic distinctions at the subspecies and species levels. Yet taxonomic assignments inevitably shape perceptions of biotic diversity, including recognition of endangered species. Case histories (involving *Ammodramus* sparrows and *Geomys* gophers) are discussed in which the data of molecular genetics revealed prior systematic errors of the two possible kinds: taxonomic recognition of groups showing little evolutionary differentiation, and lack of taxonomic recognition of phylogenetically distinct forms. In such cases, conservation efforts for "endangered species" can be misdirected with respect to the goal of protecting biological diversity.

ADDENDUM

Now, molecular genetic reappraisals of the taxonomy for threatened and endangered species are a standard part of nearly all conservation programs.

Mitochondrial gene trees and the evolutionary relationship of mallard and black ducks

Avise, J.C., C.D. Ankney, and W.S. Nelson. 1990. Mitochondrial gene trees and the evolutionary relationship of mallard and black ducks. Evolution 44:1109–1119.

ANECDOTE OR BACKDROP

Today it is universally acknowledged that a clear distinction exists between a "species phylogeny" and the multitudinous "gene trees" of which it is composed. But in 1990 this distinction between gene trees and species trees was barely on scientists' radar screen. This paper provided one early report in which a mitochondrial gene tree was analyzed and interpreted within the context of conventional phylogenetic understanding about a particular pair of extremely closely related avian species.

ABSTRACT

We assayed restriction site differences in mitochondrial (mt) DNA within and among allopatric populations of the Mallard (*Anas platyrhynchos*) and the American Black Duck (*A. rubripes*). The observed mtDNA clones grouped into two phylogenetically distinct arrays that we estimate differ by about 0.8% in nucleotide sequence. Genotypes in one clonal array were present in both species, while genotypes in the other array were seen only in Mallards. In terms of the mtDNA "gene tree", the assayed Mallards exhibit a paraphyletic relationship with respect to Black Ducks, meaning that genealogical separations among some extant haplotypes in the Mallard predate the species separation. Evidence is advanced that this pattern probably resulted from demographically based processes of lineage sorting, rather than recent secondary introgressive hybridization. However, haplotype frequencies were most similar among conspecific populations, so the Mallard and Black Ducks cluster separately in terms of a population phenogram. The results provide a clear example of the distinction between a gene tree and a population tree, and of the distinction between data analyses that view individuals versus populations as operational taxonomic units (OTUs). Overall, the mtDNA data indicate an extremely close evolutionary relationship between Mallards and Black Ducks, and in conjunction with the geographic distributions suggest that the Black Duck is a recent evolutionary derivative of a more broadly distributed Mallard–Black ancestor.

FIGURE 12.9　Mallard, *Anas platyrhynchos*.

Patterns of mitochondrial DNA and allozyme evolution in the avian genus *Ammodramus*

Zink, R.M. and J.C. Avise. 1990. Patterns of mitochondrial DNA and allozyme evolution in the avian genus Ammodramus. *Systematic Zoology 39:148–161.*

ANECDOTE OR BACKDROP

This paper was a follow-up to JCA's genetic analysis of the Seaside Sparrow (see the earlier abstract in this Chapter). It extended the mitochondrial and allozyme surveys to include all North American representatives of the secretive and cryptic complex of Ammodramus *sparrows.*

ABSTRACT

Analyses of mitochondrial (mt) DNA and allozymes were used to estimate phylogenetic patterns in the avian genus *Ammodramus*. Levels of interspecific genetic differentiation were greater than most previous estimates for other congeneric avian taxa. Phenetic and phylogenetic patterns were highly concordant for these two genetically independent data sets, suggesting a robust estimate of evolutionary history of these sparrows. However, the genetic pattern was not concordant with an estimate of variation in skeletal morphometrics produced by other authors; we suggest that ecological pressures drive convergence in skeletal morphology. Independent calibrations of mtDNA and allozyme distances suggest times of divergence that differ by a factor of two among the species assayed.

Mitochondrial DNA and avian microevolution

Avise, J.C. and R.M. Ball., Jr. 1991. Mitochondrial DNA and avian microevolution. **Acta XX Congressus Internationalis Ornithologici 1:514–524.**

ANECDOTE OR BACKDROP

This was an invited review article on early mtDNA studies in birds.

ABSTRACT

Mitochondrial DNA provides a rich source of uniparentally-inherited genetic markers especially useful in the study of matriarchal phylogeny over microevolutionary timescales. Various applications of mtDNA data to avian taxa are reviewed, including the assessments of: 1) magnitudes of intraspecific polymorphism; (2) genetic distinctions between sibling species; and (3) genetic distances and phylogenies among avian congeners. Special attention is focused on patterns of geographic differentiation within avian species, where conspecific populations have proved to exhibit a wide variety of mtDNA phylogeographic structures. The deep and geographically structured subdivisions observed in the mtDNA genealogies of some avian species probably evidence the effects of Pleistocene biogeographic separations, while the additional, shallower mtDNA subdivisions and haplotype frequency shifts distinguishing populations of most bird species probably reflect more recent restrictions on gene flow. In general, mtDNA data offer novel phylogenetic perspectives on microevolutionary processes, and allow provisional interpretation of contemporary population strucure in terms of historical demography.

ADDENDUM

Today and in recent decades, mtDNA analyses are extremely popular with ornithologists.

Matriarchal population genetic structure in an avian species with female natal philopatry

Avise, J.C., R.T. Alisauskas, W.S. Nelson, and C.D. Ankney. 1992. Matriarchal population genetic structure in an avian species with female natal philopatry. **Evolution 46:1084–1096.**

ANECDOTE OR BACKDROP

Ray Alisauskas and Dave Ankney had spent much of their scientific careers studying mating behaviors and migrational patterns in colonies of Snow Goose scattered across the species' breeding distribution in sub-Arctic regions of the Northern Hemisphere, based mostly on field observations and physical tagging studies. Here they teamed with JCA's laboratory to incorporate molecular markers into the mix. The primary intent was to determine whether or not females had always faithfully returned to their natal sites to nest (as had proved to be true for many sea turtles, as described in Chapter 9).

ABSTRACT

We employ mitochondrial (mt) DNA markers to examine the matrilineal component of population genetic structure in the snow goose, *Chen caerulescens*. From banding returns, it is known that females typically nest at their natal or prior nest site, whereas males pair with females on mixed wintering grounds and mediate considerable nuclear gene flow between geographically separate breeding colonies. Despite site philopatry documented for females, mtDNA markers show no clear distinctions between nesting populations across the species' range from Wrangel Island, USSR to Baffin Island in the eastern Canadian Arctic. Two major mtDNA clades (as well as rare haplotypes) are distributed widely and provide one of the few available examples of a phylogeographic pattern in which phylogenetic discontinuity in a gene tree exists without obvious geographic localization within a species' range. The major mtDNA clades may have differentiated in Pleistocene refugia, and colonized current nesting sites through recent range expansion via pulsed or continual low-level dispersal by females. The contrast between results of banding returns and mtDNA distributions in the snow goose raises general issues regarding population structure: direct contemporary observations on dispersal and gene flow can in some cases convey a misleading impression of phylogeographic population structure, because they fail to access the evolutionary component of population connectedness; conversely, geographic distributions of genetic markers can provide a misleading impression of contemporary dispersal and gene flow, because they retain a record of evolutionary events and past demographic parameters that may differ from those of the present. An understanding of population structure requires integration of both evolutionary (genetic) and contemporary (direct observational) perspectives.

ADDENDUM

The Snow Goose comes in two distinct and genetically controlled color phases: white and gray/blue. It was formerly thought that these two morphs might register separate species, but members of these two color-phases readily interbreed (producing offspring of either color), and both are found in substantial frequency throughout the species' geographic range. Thus, these two morphs are now generally considered conspecific, with the color phases being an intraspecific polymorphism in which the dark phase is encoded by a single dominant gene, and the white phase is homozygous for a recessive allele. Interestingly, the two mitochondrial clades discovered in this report do not conform to the two color phases, but likewise appear to constitute a straightforward molecular polymorphism within a single well-mixed taxon.

FIGURE 12.10   Snow Goose, *Chen caerulescens*.

Mitochondrial DNA phylogeographic differentiation among avian populations and the evolutionary significance of subspecies

Ball, R.M., Jr. and J.C. Avise. 1992. Mitochondrial DNA phylogeographic differentiation among avian populations and the evolutionary significance of subspecies. **The Auk** *109:626–636.*

ANECDOTE OR BACKDROP

Marty Ball was another of JCA's graduate students during the late 1980s. In addition to conducting genetic analyses (as described earlier) of the Red-winged Blackbird, he became involved in continent-wide phylogeographic surveys of several other bird species, and this work in turn led him to concerns about the evolutionary significance of traditionally described avian "subspecies."

ABSTRACT

Phylogeographic population structures revealed by restriction analyses of mitochondrial (mt) DNA were assessed within each of six avian species with continent-wide distributions in North America. The magnitude and geographic pattern of mtDNA variation differed considerably among species. The Downy Woodpecker (*Picoides pubescens*) and Mourning Dove (*Zenaida macroura*) exhibited little mtDNA polymorphism and a shallow phylogeographic structure. The Brown-headed Cowbird (*Molothrus ater*) and Song Sparrow (*Melospiza melodia*) showed somewhat higher nucleotide diversity, but no evidence of longstanding population separations. For each of these four species, evolutionary effective sizes of female populations (as estimated from mtDNA) were substantially smaller than population sizes at the present time, suggesting historical demographic constraints on the numbers of females through which mtDNA lineages have been successfully transmitted. In contrast, the Rufous-sided Towhee (*Pipilo erythrophthalmus*) and Common Yellowthroat (*Geothlypis trichas*) showed relatively deep mtDNA separations (mean nucleotide sequence divergence $p = 0.008$ and $p = 0.012$, respectively) between populations in Washinton versus those in the central and eastern states. In the case of the Rufous-sided Towhee, the mtDNA clades may correspond to morphological and behavioral differences distinguishing the western "Spotted Towhee," which was formerly recognized as a distinct species. Overall, however, most of the taxonomic subspecies currently recognized within the six assayed species were genetically very close, and showed no obvious mtDNA differences. These results raise questions concerning the population genetic and evolutionary significance of current subspecies designations in ornithology.

ADDENDUM

Partly as a result of this genetic survey, ornithologists now recognize two North American species of towhee in the "rufous-sided" complex: the Eastern Towhee (Pipilo erythrophthalmus) of the eastern United States; and the Spotted Towhee (Pipilo maculatus) in western states. More generally, geneticists have now surveyed geographic variation in many North American bird species, often finding substantial genetic differences among particular arrays of populations. In some cases, the researchers then argue that such genetic differences should be reflected in species-level or subspecies-level taxonomies; but other researchers sometimes dispute such claims. What has emerged is an oft-heated debate between the so-called taxonomic "splitters" and "lumpers". Such debates are likely to linger, because there is no universally accepted yardstick by which taxonomic assignments are recorded.

FIGURE 12.11 Downy Woodpecker, *Picoides pubescens*.

Application of genealogical-concordance principles to the taxonomy and evolutionary history of the sharp-tailed sparrow (*Ammodramus caudacutus*)

Rising, J.D. and J.C. Avise. 1993. Application of genealogical-concordance principles to the taxonomy and evolutionary history of the sharp-tailed sparrow (Ammodramus caudacutus). The Auk 110:844–856.

ANECDOTE OR BACKDROP

Jim Rising is an ornithologist well known for his longstanding work on the taxonomy and field biology of the Sharp-tailed Sparrow and related species. In this paper, he teamed with JCA to conduct refined genetic reappraisals of biogeographic conclusions he had reached previously based on morphological and behavioral comparisons in this species. The results of this study provided an early example of how molecular-genetic and traditional biological data could be fruitfully wed to yield definitive conclusions relevant to avian taxonomy and phylogeography.

ABSTRACT

We examined geographic differentiation in mitochondrial (mt) DNA and in morphometric characters among 12 populations of the Sharp-tailed Sparrow (*Ammodramus caudacutus*) representing all recognized subspecies and geographic regions. Both data sets reveal the existence of two distinct groups of populations: a northern group from the Canadian maritime provinces and Maine, the St. Lawrence Valley, Hudson Bay lowlands, and interior prairies; and a southern group from along the Atlantic coast north to southern Maine. In one sample from southern Maine, both forms co-occur and about 40% of the individuals appear to be of hybrid ancestry. Recently, principles of genealogical concordance have been proposed as a conceptual basis for recognition of biological taxa. Here we provide an empirical application of these principles in the context of the observed concordance between the mtDNA phylogenetic partition and the subdivisions evidenced by morphological (and behavioral) attributes in the Sharp-tailed Sparrow complex. We recommend that two subspecies of *A. caudacutus* be recognized: one (*A. c. nelsoni*) to encompass the northern populations (formerly *A. c. nelsoni*, *A. c. alterus*, and *A. c. subvirgatus*); and the other (*A. c. caudacutus*) to encompass the southern populations (formerly *A. c. caudacutus* and *A. c. diversus*). By taxonomically formalizing what appears to be a fundamental phylogenetic partition between these Sharp-tailed Sparrow populations, study of the biogeographic history, reproductive relationships, and management of the forms will be facilitated.

ADDENDUM

Partly as a result of this genetic survey, ornithologists now recognize two North American species of sparrow in the "sharp-tailed" complex: Nelson's Sharp-tailed Sparrow (Ammodramus nelsoni) to the north; and the Saltmarsh Sharp-tailed Sparrow (Ammodramus caudacutus) to the south. Thus, this could be considered a case in which taxonomic "splitters" have won the debate, but with good reason. More and more systematists are coming around to the view that concordant support from multiple genetic or other traits is both necessary and sufficient evidence for the taxonomic recognition of particular biological entities in question.

FIGURE 12.12   Sharp-tailed Sparrow, *Ammodramus caudacutus*.

Three fundamental contributions of molecular genetics to avian ecology and evolution

Avise, J.C. 1996. Three fundamental contributions of molecular genetics to avian ecology and evolution. Ibis 138:16–25.

ANECDOTE OR BACKDROP

This was an invited review paper for an ornithological congress.

ABSTRACT

Studies in molecular genetics are having revisionary impact in at least three broad areas of avian ecology and evolution: mating systems; geographic population structure and gene flow; and phylogenetic relationships among species and higher taxa. With regard to mating systems, genetic analyses of maternity and paternity have revealed unexpectedly high frequencies of extra-pair fertilization and intraspecific brood parasitism in numerous avian species (including those thought to be socially monogamous), and these discoveries are prompting a fundamental reshaping of mating system theory for birds. With regard to genetic structure, molecular markers have uncovered a great variety of depths and patterns in the phylogeographic histories of conspecific populations, and these findings provide novel perspectives on historical gene flow regimes and species concepts. With regard to evolutionary relationships among higher avian taxa, molecular findings have suggested several phylogenetic realignments, thus prompting renewed interest in the cross-comparative aspects of molecular and morphological evolution as well as of alternative procedures for molecular analysis.

ADDENDUM

All of these conclusions could be restated again today, but in a much amplified form and with countless more empirical examples.

Pleistocene phylogeographic effects on avian populations and the speciation process

Avise, J.C. and D. Walker. 1998. Pleistocene phylogeographic effects on avian populations and the speciation process. **Proceedings of the Royal Society of London B 265:457–463.**

ANECDOTE OR BACKDROP

Conventional wisdom in ornithology was that environmental changes associated with glacial cycles of the Pleistocene Epoch precipitated many of the speciation events between closely related extant bird species. In an important and provocative publication, Klicka and Zink (Science 277:1666–1669) challenged this paradigm by claiming—based on mtDNA evidence—that most such avian speciations actually predated the recent Ice Ages. In turn, our next paper challenged the Klicka-Zink scenario. Much of the debate centers on and thereby highlights the key distinction between gene trees and species trees.

ABSTRACT

Pleistocene biogeographic events have traditionally been ascribed a major role in promoting speciations and in sculpting the present-day diversity and distributions of vertebrate taxa. However, this paradigm has recently come under challenge from a review of interspecific mtDNA genetic distances in birds: most sister-species separations dated to the Pliocene. Here we summarize the literature on intraspecific mtDNA phylogeographic patterns in birds and reinterpret the molecular evidence bearing on Pleistocene influences. At least 37 of the 63 avian species surveyed (59%) are sundered into recognizable phylogeographic units, and 28 of these separations (76%) trace to the Pleistocene. Furthermore, the use of phylogroup separation times within species as minimum estimates of "speciation durations" also indicates that many protracted speciations, considered individually, probably extended through time from Pliocene origins to Pleistocene completions. When avian speciation is viewed properly as an extended temporal process rather than as a point event, Pleistocene conditions appear to have played an active role both in initiating major phylogeographic separations within species, and in completing speciations that had been inaugurated earlier. Whether the Pleistocene was exceptional in these regards compared with other geological times remains to be determined.

ADDENDUM

Such appraisals need to be expanded to many additional avian taxa, including those inhabiting tropical and subtropical regions of the world (which are relatively poorly represented in existing genetic surveys).

Matrilineal history of the endangered Cape Sable seaside sparrow inferred from mitochondrial DNA polymorphism

Nelson, W.S., T. Dean, and J.C. Avise. 2000. Matrilineal history of the endangered Cape Sable seaside sparrow inferred from mitochondrial DNA polymorphism. **Molecular Ecology** *9:809–813.*

ANECDOTE OR BACKDROP

Another endangered "species" in the Seaside Sparrow complex is the Cape Sable race of extreme southern Florida. The focal specimen used in our genetic analysis came from what was undoubtedly the most unusual sampling protocol ever employed in JCA's lab for any animal species. Namely, the specimen in question had been captured and fitted with a small radio transmitter before being released back into nature. A few days later, the corpse of the unlucky bird was radio-tracked to and then retrieved from the stomach of a wild snake who had eaten it! Fortunately for us, the bird's tissues still remained suitable for our subsequent molecular assays.

ABSTRACT

Restriction analyses were conducted on mitochondrial (mt) DNA amplified by long-PCR from an endangered bird, the Cape Sable Seaside Sparrow. The first of several successful mtDNA amplifications was accomplished using the partially digested tissues of a transmitter-monitored bird retrieved from the gut of a snake. As many as 91 mtDNA restriction fragments produced by 18 endonucleases were compared in this and four other Cape Sable specimens against mtDNA similarly amplified by long-PCR from other taxonomic forms in the seaside sparrow complex. Results indicate that the Cape Sable birds belong to an Atlantic matrilineal clade, and are highly divergent from other seaside sparrows along the Gulf of Mexico.

ADDENDUM

The Cape Sable Seaside Sparrow (Ammodramus maritimus mirabilis) is still considered endangered due to habitat loss and the threat of hurricanes. Its range is confined mostly to prairies in Everglades National Park and Big Cypress Swamp. Recent lawsuits by environmentalists have focused on water-management practices (by the U.S. Army Corps of Engineers) that affect the breeding habitat of this subspecies.

Phylogeography of colonially nesting seabirds, with special reference to global matrilineal patterns in the sooty tern (*Sterna fuscata*)

Avise, J.C., W.S. Nelson, B.W. Bowen, and D. Walker. 2000. Phylogeography of colonially nesting seabirds, with special reference to global matrilineal patterns in the sooty tern (Sterna fuscata). Molecular Ecology 9:1783–1792.

ANECDOTE OR BACKDROP

The Sooty Tern is an avian analogue of the Green Turtle (see Chapter 9) in several regards: both species are long-lived with delayed breeding; both are circumtropically distributed and tend to nest in discrete rookeries (often sharing the same sandy islands); and individuals in both species tend to be fidelic to natal sites despite traveling tremendous distances at non-nesting stages of the life cycle. Do these many similarities in population demography and lifestyle translate into similar population genetic architectures in these otherwise unrelated pelagic mariners? This paper addresses the issue, on a global phylogeographic scale.

ABSTRACT

Sooty tern (*Sterna fuscata*) rookeries are scattered throughout the tropical oceans. When not nesting, individuals wander great distances across open seas, but, like many other seabirds, they tend to be site-faithful to nesting locales in successive years. Here we examine the matrilineal history of sooty terns on a global scale. Assayed colonies within an ocean are poorly differentiated in mitochondrial DNA sequence, a result indicating tight historical ties. However, a shallow genealogical partition distinguishes Atlantic from Indo-Pacific rookeries. Phylogeographic patterns in the sooty tern are compared to those in other colonially nesting seabirds, as well as in the green turtle (*Chelonia mydas*), an analogue of tropical seabirds in some salient aspects of natural history. Phylogeographic structure within an ocean is normally weak in seabirds, unlike the pronounced matrilineal structure in green turtles. However, the phylogeographic partition between Atlantic and Indo-Pacific rookeries in sooty terns mirrors, albeit in shallower evolutionary time, the major matrilineal subdivision in green turtles. Thus, global geology has apparently influenced historical gene movements in these two circumtropical species.

ADDENDUM

The Sooty Tern is one of the world's most abundant and widely distributed avian species. It is colloquially called the "wideawake" tern, or simply the wideawake, a name that stems from the incessant calls issued by these birds on their raucous breeding colonies.

FIGURE 12.13 Sooty Tern, *Sterna fuscata*.

13

Rodents

INTRODUCTION

When JCA began graduate school at the University of Texas in 1970, the laboratory of his thesis advisor [Robert K. Selander (RKS)] was preoccupied with applying newly discovered allozyme methods to rodents and various other organisms. Multilocus protein-electrophoretic techniques had just been introduced to population biology, and RKS's laboratory was busy extending this exciting new method to creatures other than standard fruit flies and humans. One taxonomic group that was a focus of the RKS lab involved mice in the speciose American genus *Peromyscus*. JCA soon became involved with these projects, and, although not part of his own thesis, he eventually helped collect the mice (in Mexico and across the southwestern United States), scored the starch gels that the RKS lab was churning out at a pace of about 20 per day, analyzed the resulting genetic data, and wrote several papers on these creatures, some of which are abstracted here. These were among the first protein-electrophoretic surveys of nonhuman vertebrates, and they were truly exploratory in their intellectual nature. For example, at that time it was still uncertain whether (and if so how) these newfangled molecular techniques might contribute to population biology, ecology, and systematics. Few natural historians (JCA included) had much prior knowledge or empirical experience with genetics, so dealing with the kinds of molecular information provided by protein electrophoresis was truly like entering a brave new world.

Nearly a decade later, JCA would return to his familiar rodents, this time in the context of exploring how another molecular technique—restriction enzyme analysis of mitochondrial DNA—might again revolutionize the neophyte fields of molecular ecology and molecular evolution.

Biochemical polymorphism and systematics in the genus *Peromyscus*. V. Insular and mainland species of the subgenus *Haplomylomys*

Avise, J.C., M.H. Smith, R.K. Selander, T.E. Lawlor, and P.R. Ramsey. 1974. Biochemical polymorphism and systematics in the genus Peromyscus. *V. Insular and mainland species of the subgenus* Haplomylomys. *Systematic Zoology 23:226–238.*

ANECDOTE OR BACKDROP

After graduating with an M.A. degree from the University of Texas in 1971, JCA was hired by Dr. Michael Smith to set up and run a protein-electrophoretic laboratory at the Savannah River Ecology Laboratory (SREL) near Aiken, SC. There, for two wonderful years, Mike gave JCA free rein to conduct allozyme analyses on almost any creatures and topics that piqued his curiosity. This paper registers one such study—of spatial population structure and evolutionary relationships in several species of Peromyscus *mice native to western North America and islands in the Gulf of California. This was among several early studies from SREL that began to convince JCA of the power of allozyme methods for revealing genetic relationships within and among closely related species. This was the fifth in a series of papers on native* Peromyscus *mice from Robert Selander's lab (where Michael Smith had taken a sabbatical leave), but it was the first one on which JCA took the lead role. This entire series of papers was quite ahead of its time, and blazed the protein-electrophoretic pathway for many other researchers in the decade that followed.*

ABSTRACT

We have examined allozyme variation at 25 loci in nine species of *Peromyscus* inhabiting the southwestern United States, Sonora, Baja California, and islands in the Gulf of California. Four previously studied species of *Peromyscus* are also included in a dendrogram formed by cluster analysis of genic similarity coefficients. Mainland populations currently assigned to *P. eremicus* represent two distinctive genetic types: an eastern form in Nevada, New Mexico, Arizona, Texas, and Sonora, and a western form in southern California and Baja that may have been separated originally by the Gulf of California embayment during the Pleistocene. *Peromyscus merriami* is genetically distinct from sympatric *P. eremicus* populations of the eastern type, although it falls within the range of genetic variation found between eastern and western *P. eremicus* forms. The insular endemics *P. guardia*, *P. interparietalis*, and *P. dickeyi*, and two insular subspecies of *P. eremicus*, are similar to the western *P. eremicus* type on Baja and probably share a recent common ancestor. Populations on shallow-water islands near Baja are more similar to mainland populations than are those on deep-water islands. Other insular species, *P. caniceps* and *P. stephani*, are genetically more distinct from *P. eremicus* and may have closer relationships with other mainland species. *Peromyscus sejugis* is closely related to *P. polionotus* (*maniculatus* species group) and probably evolved from *P. maniculatus*. There is considerable variation in level of genic heterozygosity among mainland populations, although the mean of 6 percent is consistent with "normal" heterozygosity estimates for other vertebrates. All insular populations have low variability, averaging less than 1 percent of loci in heterozygous state, presumably as a consequence of random drift in small populations.

ADDENDUM

At the time this study was published, Michael Smith had recently become director of the SREL. The two years that JCA spent at SREL constituted alternative service into which the author had been drafted during the era of the Vietnam war.

FIGURE 13.1 Cactus mouse, *Peromyscus eremicus*.

Biochemical polymorphism and systematics in the genus *Peromyscus*. VI. Systematic relationships in the *boylii* species group

Avise, J.C., M.H. Smith, and R.K. Selander. 1974. Biochemical polymorphism and systematics in the genus Peromyscus. *VI. Systematic relationships in the* boylii *species group.* Journal of Mammalogy *55:751–763.*

ANECDOTE OR BACKDROP

This allozyme-based paper was another in the now-classic Peromyscus *series of manuscripts during the early 1970s.*

ABSTRACT

An analysis of electrophoretic variation in proteins encoded by 21 genetic loci in 275 individuals belonging to the *Peromyscus boylii* species group yielded the following systematic conclusions: populations of *P. boylii rowleyi* and *P. b. levipes* in four Mexican states and four states in the southwestern United States, separated by up to 3,000 kilometers, share common alleles at virtually all loci and thus show no evidence of representing more than one species; *P. b. attwateri* from Arkansas differs from *P. boylii* in allelic composition at several loci, and in all probability represents a distinct species, as suggested by other authors on the basis of morphology and karyotype; and *P. stephani* from the Gulf of California is very similar genetically to *P. boylii*, and, on the basis of this and other evidence, should be removed from the subgenus *Haplomylomys* and placed in the *boylii* species group of the subgenus *Peromyscus*. Populations referable to *P. pectoralis* from Ciudad Victoria, Mexico, Ranger, Texas, and Big Bend, Texas, show considerable allelic differences from one another but form a single cluster in a dendrogram of biochemical similarities. The nature of the allelic differences suggests that two or more species currently are classified as *P. pectoralis*.

ADDENDUM

More than 50 species of Peromyscus *currently are recognized, so plenty of room remains for further genetic analyses of this species-rich group of New World rodents.*

Biochemical polymorphism and systematics in the genus *Peromyscus*. VII. Geographic differentiation in members of the *truei* and *maniculatus* species groups

Avise, J.C., M.H. Smith, and R.K. Selander. 1979. Biochemical polymorphism and systematics in the genus Peromyscus. *VII. Geographic differentiation in members of the* truei *and* maniculatus *species groups.* Journal of Mammalogy *60:177–192.*

ANECDOTE OR BACKDROP

The Deer Mouse Peromyscus maniculatus *is probably the most abundant and widespread mammal on the North American continent, with extensive geographic variation in morphology and many described races. This article added the Deer Mouse and several other native mice species to the growing catalogue of* Peromyscus *taxa for which genetic data were collected and analyzed very early in the protein-electrophoretic era.*

ABSTRACT

Allozymic variation at 22 gene loci in populations of *Peromyscus truei*, *P. difficilis*, *P. melanotis*, and *P. maniculatus* is used to examine patterns of geographic differentiation. Samples of *P. maniculatus* collected throughout most of its immense range, from southern Mexico to northern Canada and from the Pacific to the Atlantic coasts, are very similar in allelic composition. Evidence from morphology, chromosomal configurations, ecology, and patterns of reproductive isolation argue that contemporary gene flow among *P. maniculatus* populations is not sufficient to account for the similarity in their allelic configurations. The macrogeographic conservatism in level of genic divergence conceivably results from a relatively recent separation of populations, coupled with a genetic inertia resulting from a selected cohesion of the genome. Nonetheless, significant intersample heterogeneity of allele frequencies at polymorphic loci may certainly result from stochastic effects. The relative geographic uniformity in *P. maniculatus* contrasts somewhat with the geographic differences observed in *P. truei* and in *P. difficilis*. Members of the *truei* and *maniculatus* species groups are added to a biochemical dendrogram, which now includes 20 named species of *Peromyscus*.

FIGURE 13.2   Deer mouse, *Peromyscus maniculatus*.

The use of restriction endonucleases to measure mitochondrial DNA sequence relatedness in natural populations. I. Population structure and evolution in the genus *Peromyscus*

Avise, J.C., R.A. Lansman, and R.O. Shade. 1979. The use of restriction endonucleases to measure mitochondrial DNA sequence relatedness in natural populations. I. Population structure and evolution in the genus Peromyscus. *Genetics 92:279–295.*

ANECDOTE OR BACKDROP

This was the first of several early papers from JCA's laboratory that inaugurated mitochondrial DNA as a powerful new molecular system to genetically compare populations and species in nature.

ABSTRACT

We introduce to natural population analysis a molecular technique that involves the use of restriction endonucleases to compare mitochondrial (mt) DNA sequences. We have examined the fragment patterns produced by six restriction endonucleases acting upon mtDNA isolated from 23 samples of three species of the rodent genus *Peromyscus*. Our observations confirm the following conclusions derived from previous experiments with laboratory animals: (i) mtDNA within an individual appears homogeneous and (ii) at least the majority of mtDNA present in an individual is inherited from the female parent. Our experiments demonstrate for the first time that there is detectable heterogeneity in mtDNA sequences within and among natural geographic populations of a species and that this heterogeneity can readily be used to estimate relatedness between individuals and populations. Individuals collected within a single locale show less than 0.5 percent sequence divergence, while those collected from conspecific populations separated by 50–500 miles differ by approximately 1.5%. The mtDNAs of the closely related sibling species *P. polionotus* and *P. maniculatus* differ from each other by 13–17%; nonsibling species differ by more than 20%. Qualitative and quantitative approaches to analysis of digestion patterns are suggested. The results indicate that restriction analysis of mtDNA may become the most sensitive and powerful technique yet available for reconstructing evolutionary relationships among conspecific organisms.

ADDENDUM

This paper has become rather famous and has been widely cited because it was the first substantive example of the beauty and utility of mitochondrial DNA for assessing genetic (matrilineal) relationships among conspecific populations and closely related species in nature. From the 1980s to the present day, mtDNA has remained a powerful workhorse for a wide range of micro-evolutionary applications in molecular ecology and evolution (although especially in recent years direct sequencing of mitochondrial genomes has supplanted earlier assay methods that employed restriction endonucleases).

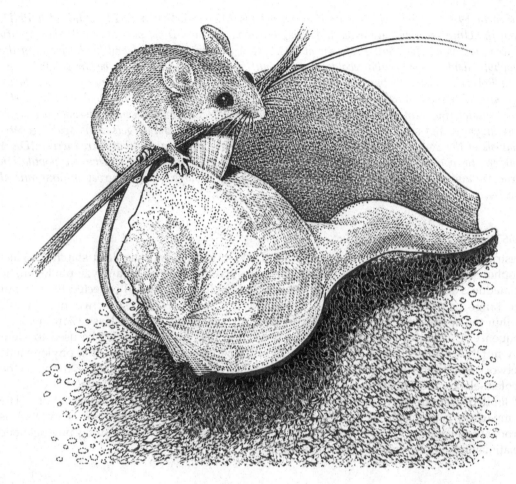

FIGURE 13.3   Beach mouse, *Peromyscus polionotus*.

Mitochondrial DNA clones and matriarchal phylogeny within and among geographic populations of the pocket gopher, *Geomys pinetis*

Avise, J.C., C. Giblin-Davidson, J. Laerm, J.C. Patton, and R.A. Lansman. 1979. Mitochondrial DNA clones and matriarchal phylogeny within and among geographic populations of the pocket gopher, Geomys pinetis. Proceedings of the National Academy of Sciences USA 76:6694–6698.

ANECDOTE OR BACKDROP

This paper was arguably the single most important publication of JCA's career. To appreciate its genesis, we can fast-forward across several years, temporarily bypassing the allozyme era of the early 1970s during which time JCA earned his M.A., worked at SREL, obtained a Ph.D. from the University of California at Davis, and took a tenure-track position at the University of Georgia. As an Assistant Professor at UGA, JCA was anxious to expand his horizons in the neophyte fields of molecular ecology and evolution. Serendipitously, he began collaborating with Robert Lansman, a biochemist who happened to work on the physiology of mitochondria. Bob and JCA soon found themselves purifying mitochondrial (mt) DNA from various vertebrate species (including Peromyscus*), digesting the molecules with recently discovered restriction enzymes, and separating the fragments on agarose or acrylamide gels. This spatio-genetic analysis of the Southeastern Pocket Gopher was one of the first papers that introduced mtDNA markers to population biology. It also pioneered a genealogical perspective on population structure and thereby in effect was the first substantial phylogeographic survey on any animal species.*

ABSTRACT

Restriction endonuclease assay of mitochondrial (mt) DNA and standard starch-gel electrophoresis of proteins encoded by nuclear genes have been used to analyze phylogenetic relatedness among a large number of pocket gophers (*Geomys pinetis*) collected throughout the range of the species. The restriction analysis clearly distinguishes two populations within the species, an eastern and a western form, which differ by at least 3% in mtDNA sequence. Qualitative comparisons of the restriction phenotypes can also be used to identify mtDNA "clones" within each form. The mtDNA clones interconnect in a phylogenetic network that represents an estimate of matriarchal phylogeny for *G. pinetis*. Although the protein-electrophoretic data also differentiate the eastern and western forms, the data are of limited usefulness in establishing relationships among more local subpopulations. The comparison between these two data sets suggests that restriction analysis of mtDNA is probably unequaled by other techniques currently available for determining phylogenetic relationships among conspecific organisms.

ADDENDUM

The Southeastern Pocket Gopher is one of nine currently recognized species of pocket gopher native to the eastern and southern United States. These species get their common name from their pocket-like cheek pouches. At the time this genetic survey was conducted, we had to purify mtDNA from fresh (not frozen) large liver samples using time-consuming Cs-Cl gradient centrifugation, followed by restrictions digests of the closed-circular mtDNA molecules. Although these analytical procedures were extremely laborious, in retrospect they did have the advantage of allowing us to avoid the potential complications of numts (mitochondrial pseudogenes housed in the cell's nuclear genome). Such numts were later found to be quite common in many animal species, sometimes compromising conclusions that could otherwise be drawn about the presumed matrilineal inheritance of mtDNA markers.

FIGURE 13.4   Southeastern Pocket Gopher, *Geomys pinetis*.

The use of restriction endonucleases to measure mitochondrial DNA sequence relatedness in natural populations. III. Techniques and potential applications

Lansman, R.A., R.O. Shade, J.F. Shapira, and J.C. Avise. 1981. The use of restriction endonucleases to measure mitochondrial DNA sequence relatedness in natural populations. III. Techniques and potential applications. Journal of Molecular Evolution 17:214–226.

ANECDOTE OR BACKDROP

This article detailed newly developed laboratory protocols and applications for mitochondrial DNA surveys using restriction endonucleases.

ABSTRACT

Restriction endonucleases and agarose gel electrophoresis have been used to demonstrate extensive nucleotide diversity in mitochondrial (mt) DNA within and between conspecific populations of rodents. Cleavage of mtDNA samples with a relatively small number of endonucleases provides information (concerning the phylogenetic relatedness of individual organisms) that cannot now be readily obtained by any other type of molecular analysis. This information is qualitatively different from that available from the study of nuclear genes or gene products because the mitochondrial genome is inherited intact from the female parent and is not altered by recombination or meiotic segregation. The techniques are described in detail in an effort to make restriction analysis of mtDNA available to biologists who may be unfamiliar with current DNA technology.

ADDENDUM

Probably because it was a "techniques paper," this publication is quite widely cited in the scientific literature.

Genetic divergence between rodent species assessed by using two-dimensional electrophoresis

Aquadro, C.F. and J.C. Avise. 1981. Genetic divergence between rodent species assessed by using two-dimensional electrophoresis. **Proceedings of the National Academy of Sciences USA** *78:3784–3788.*

ANECDOTE OR BACKDROP

Graduate student Charles (Chip) Aquadro had a penchant for experimenting with almost any new molecular genetic technique as soon as it had been introduced. Here he used a novel laboratory procedure—two-dimensional electrophoresis—to explore whether this methodology might find highly useful applications in surveys of natural populations and species. The broad answer was "not so much."

ABSTRACT

O'Farrell's technique of two-dimensional gel electrophoresis (2-DGE) has previously been applied to the study of intrapopulation genetic variation. This approach assays a larger, and in part nonoverlapping, cohort of protein encoding loci compared to conventional one-dimensional starch-gel electrophoretic procedures (SGE) and has revealed substantially lower levels of mean heterozygosity. Here we extend this approach to analyze levels and patterns of genetic differentiation between species. We have used 2-DGE to compare an average of 189 polypeptides between six species of wild mice representing levels of evolutionary divergence ranging from different subspecies to different families. The magnitude of protein divergence estimated by 2-DGE was on the average only about one-half that predicted by SGE. This discrepancy may result from differences in sensitivities between the techniques or differences in the mean level of variation and divergence between the sets of loci assayed by the two methods. Nonetheless, the ranking of genetic distances by 2-DGE was identical to that by SGE. Thus, these results support the use of the simpler SGE techniques to estimate relative levels of genetic divergence.

ADDENDUM

Two-dimensional gel electrophoresis was just a passing fancy for the broader field of evolutionary genetics, lasting only a few years and providing little useful genetic information.

Genetic determination of the status of an endangered species of pocket gopher in Georgia

Laerm, J., J.C. Avise, J.C. Patton, and R.A. Lansman, 1982. Genetic determination of the status of an endangered species of pocket gopher in Georgia. **Journal of Wildlife Management** *46:513–518.*

ANECDOTE OR BACKDROP

The U.S. Endangered Species Act (ESA) of 1973 grants formal protection to rare and endangered species and is probably our country's strongest legal framework for protecting native species facing extinction. But what exactly is an endangered species? The ESA defines an endangered species as a species, a subspecies, or a "distinct population segment" that is at risk of extinction throughout all or a significant portion of its range. Traditionally, taxonomic assignments were taken at face value, and a primary task for wildlife biologists was to assess the numerical abundance and distribution of each taxon that was included or nominated for the endangered species list. This article was perhaps the first multifaceted genetic analysis of any endangered species in which the primary issue was not the species' demographic status in terms of abundance and range, but rather the species' taxonomic status with respect to its genetic distinctiveness.

ABSTRACT

In the southeastern United States, *Geomys* pocket gophers are represented by two extant taxonomic species: *G. colonus* and *G. pinetis*. The former was described in 1898 but remained nearly unnoticed and unstudied until 1967 when a population within the historical range of *colonus* was "rediscovered." The current population referable to *"colonus"* encompasses about 16 km^2 of coastal plain in Camden County, GA and is presently listed as an endangered species by the State of Georgia. Here we have compared many body and cranial measurements, as well as genetic data from allozymes and mitochondrial DNA, to compare *colonus* against geographic populations of the far more common species *G. pinetis*. Considering all available evidence, we reject the hypothesis that the gopher population presently recognized as *Geomys colonus* differs significantly in genetic composition from surrounding populations of *G. pinetis*. This conclusion is not an artifact of poor sensitivity in the genetic assays employed, as these same techniques indicated considerable genetic differences among geographic populations of *G. pinetis*. Therefore, *"colonus"* appears to represent no more than a slightly differentiated local population of the eastern form of *G. pinetis* and should be synonymized with it.

ADDENDUM

The colonial pocket gopher is no longer recognized as a valid taxonomic or biological entity.

An assessment of "hidden" heterogeneity within electromorphs at three enzyme loci in deer mice

Aquadro, C.F. and J.C. Avise. 1982. An assessment of "hidden" heterogeneity within electromorphs at three enzyme loci in deer mice. **Genetics 102:269–284.**

ANECDOTE OR BACKDROP

I mentioned earlier that, as a student, Chip Aquadro loved to tinker with new molecular techniques in the laboratory. This article provides another illustration of his instinct for such genetic exploration.

ABSTRACT

Allelic heterogeneity within protein electromorphs at three loci was examined in populations of deer mice (*Peromyscus maniculatus*) collected from five localities across North America. We used a variety of electrophoretic techniques (including several starch and acrylamide conditions, gel-sieving, and isoelectric focusing), plus heat denaturation. Of particular interest was the supernatant glutamate oxalate transaminase system (GOT-1), which under standard electrophoretic conditions had been shown to exhibit basically a two-allele polymorphism throughout the range of *maniculatus*. The use of all of the above techniques failed to uncover any additional variation for GOT-1 in these populations. Similarly, no new scorable variation was resolved at the essentially monomorphic malate dehydrogenase-1 locus by additional conditions of electrophoresis. In marked contrast to the results for the above two enzymes, the use of multiple conditions of electrophoresis resolved the eight standard condition electromorphs of esterase-1 into a total of 23 variants showing strong geographic differentiation in frequency. These 23 electromorphs were further divided into a total of 35 variants by thermal stability studies. However, the allelic nature of the thermal stability esterase variants remains to be documented. The results of this study, taken together with the remarkable geographic heterogeneity for this species in ecology, morphology, karyotype, and mitochondrial DNA sequence, suggest that some form of balancing selection may be acting to maintain the GOT-1 polymorphism.

ADDENDUM

The various techniques employed in this article were also passing short-lived fads in the history of molecular evolutionary genetics.

Extensive genetic variation in mitochondrial DNA's among geographic populations of the deer mouse, *Peromyscus maniculatus*

Lansman, R.A., J.C. Avise, C.F. Aquadro, J.F. Shapira, and S.W. Daniel. 1983. Extensive genetic variation in mitochondrial DNA's among geographic populations of the deer mouse, Peromyscus maniculatus. *Evolution 37:1–16.*

ANECDOTE OR BACKDROP

About one decade after he first worked on the Deer Mouse using allozyme methods (see above), JCA returned to this widespread rodent as a part of his broader efforts to explore the potential of mitochondrial DNA for geographic surveys of animal populations. The Deer Mouse proved to be a fine biological subject for such analyses, because the species revealed abundant mtDNA genetic variation that had an obvious geographic orientation.

ABSTRACT

We present an extensive analysis of mitochondrial (mt) DNA nucleotide sequence variation within the rodent *Peromyscus maniculatus*. Using techniques that allow assay of mtDNA from individual animals, we have mapped variable and conserved cleavage sites for 8 restriction endonucleases in mtDNA prepared from 135 animals collected throughout

North America. The data provide information on two different but interrelated aspects of mtDNA evolution: (i) the nature of evolutionary change in the molecule itself and (ii) the implications of these molecular changes for estimating matriarchal relationships within *P. maniculatus*. Our data support the hypothesis that mammalian mtDNA evolves primarily by the accumulation of single base substitutions which occur in most of the coding regions of the genome. Sequence divergence between *P. maniculatus* mtDNA samples can be as high as 7 percent. The distributions of cleavage sites and the magnitudes of sequence divergence among samples are strongly related to the geographic sources of the collections. The data clearly distinguish five major genetic assemblages within *P. maniculatus*, as well as extensive clonal diversity within each of those assemblages. Convergent or parallel sequence alterations are not uncommon. Phylogenies derived from mtDNA restriction analyses are not generally concordant with those previously estimated from morphological considerations.

Mitochondrial DNA differentiation during the speciation process in *Peromyscus*

Avise, J.C., J.F. Shapira, S.W. Daniel, C.F. Aquadro, and R.A. Lansman. 1983. Mitochondrial DNA differentiation during the speciation process in Peromyscus. **Molecular Biology and Evolution** *1:38–56.*

ANECDOTE OR BACKDROP

In 1983, the fundamental distinction between a species and a gene tree was just beginning to percolate through the consciousness of molecular evolutionists. This article was one of the first to empirically address how a gene tree (in this case from mitochondrial DNA) might become partitioned during the course of biological speciation between closely related taxa. It turns out that historical population demography plays a key role in this genealogical process.

ABSTRACT

We address the possible significance of biological speciation to the magnitude and pattern of divergence of asexually transmitted characters in bisexual species. The empirical data for this report consist of restriction endonuclease site variability in maternally transmitted mitochondrial (mt) DNA isolated from 82 samples of *Peromyscus polionotus* and *P. leucopus* collected from major portions of the respective species' ranges. Data are analyzed together with previously published information on *P. maniculatus*, a sibling species to *P. polionotus*. Maps of restriction sites indicate that all of the variation observed can be reasonably attributed to base substitutions leading to loss or gain of particular restriction sites. Magnitude of mtDNA sequence divergence within *polionotus* (maximum = 2%) is roughly comparable to that observed within any of five previously identified mtDNA assemblages in *maniculatus*. Sequence divergence within *leucopus* (maximum = 4%) is somewhat greater than that within *polionotus*. Consideration of probable evolutionary links among mtDNA restriction site maps allowed estimation of matriarchal phylogenies within *polionotus* and *leucopus*. Clustering algorithms and qualitative Wagner procedures were used to generate phenograms and parsimony networks, respectively, for the between-species comparisons.

Three simple graphical models are presented to illustrate some conceivable relationships of mtDNA differentiation to speciation. In theoretical case I, each of two reproductively defined species (A and B) is monophyletic in matriarchal genealogy; the common female ancestor can either predate or postdate the speciation. In case II, neither species is monophyletic in matriarchal genotype. In case III, species B is monophyletic but forms a subclade within A which is thus paraphyletic with respect to B. The empirical results for mtDNA in *maniculatus* and *polionotus* appear to conform closely to case III. These theoretical and empirical considerations raise a number of questions about the general relationship of the speciation process to the evolution of uniparentally transmitted traits. Some of these considerations are presented, and it is suggested that the distribution patterns of mtDNA sequence variation within and among extant species should be of considerable relevance to the particular demographies of speciation.

ADDENDUM

Peromyscus maniculatus *and* P. polionotus *continue to be recognized as quintessential examples of closely related sibling species.*

Microgeographic lineage analysis by mitochondrial genotype: variation in the cotton rat (*Sigmodon hispidus*)

*Kessler, L.G. and J.C. Avise. 1985. Microgeographic lineage analysis by mitochondrial genotype: variation in the cotton rat (*Sigmodon hispidus*). Evolution 39:831–837.*

ANECDOTE OR BACKDROP

The Cotton Rat looks like a regular house rat but with frizzy hair (as if the animal had been under a blow dryer). The native species is common throughout the southeastern United States, is easy to trap, and thus for a brief time became an object for empirical genetic research in JCA's lab. This study extended mtDNA analyses of rodent families down to a microgeographic scale.

ABSTRACT

We evaluate the use of mitochondrial (mt) DNA genetic markers to describe population structure and matrilineal kinship on a microgeographic scale. An analysis of restriction site variation in 134 cotton rats (*Sigmodon hispidus*) collected from a 3.2-hectare field revealed significant spatial and temporal heterogeneity in frequencies of mtDNA genotypes among nest sites. Inspection of particular genotypes provided additional information about minimum numbers of female lineages (family units) per nest site, and the possible matrilineal affiliations of individuals. However, since shared genotypes are not necessarily synapomorphs having arisen within the study area, conclusions about dispersal must remain reserved. Study of the maternally inherited mtDNA genome offers novel perspectives on the meaning of microgeographic population structure.

FIGURE 13.5   Cotton rat, *Sigmodon hispidus.*

Estimation of single generation migration distances from geographic variation in animal mitochondrial DNA

Neigel, J.E., R.M. Ball., Jr., and J.C. Avise. 1991. Estimation of single generation migration distances from geographic variation in animal mitochondrial DNA. Evolution 45:423–432.

ANECDOTE OR BACKDROP

There seems often to be a bit of professional jealousy between outdoor biologists (such as field ecologists) and laboratory biologists (such as many molecular geneticists). These tensions probably reflect many factors such as funding priorities at granting agencies and inherently different approaches to science. In any event, the most exciting advances sometimes come when conceptual or empirical bridges are built between the laboratory and field. This article provides one such example. It shows how a new kind of phylogeographic analysis of mtDNA data might be reconciled with direct field observations to evaluate the dispersal capacity of small mammals on a per-generation basis.

ABSTRACT

A new approach is introduced for the analysis of dispersal from the geographic distributions of mtDNA lineages. The method is based on the expected spatial distributions of lineages arising under a multigeneration random walk process. Unlike previous methods based on the predicted equilibria between genetic drift and gene flow, this approach is appropriate for nonequilibrium conditions and yields an estimate of dispersal distance rather than dispersal rate. The theoretical basis for this method is examined, and an analysis of mtDNA restriction site data for *Peromyscus maniculatus* is presented as an example of how this approach can be applied to empirical data.

Application of a random walk model to geographic distributions of animal mitochondrial DNA variation

Neigel, J.E. and J.C. Avise. 1993. Application of a random walk model to geographic distributions of animal mitochondrial DNA variation. **Genetics** *135:1209–1220.*

ANECDOTE OR BACKDROP

This conceptually novel study provided a logical follow-up to the prior paper abstracted above. It again entailed analyses of mitochondrial data from rodents.

ABSTRACT

In rapidly evolving molecules, such as animal mitochondrial DNA, mutations that delineate specific lineages may not be dispersed at sufficient rates to attain an equilibrium between genetic drift and gene flow. Here we predict conditions that lead to nonequilibrium geographic distributions of mtDNA lineages, test the robustness of these predictions, and examine mtDNA data sets for consistency with our model. Under a simple isolation-by-distance model, the variance of an mtDNA lineage's geographic distribution is expected to be proportional to its age. Simulation results indicated that this relationship is fairly robust. Analysis of mtDNA data from natural populations (of *Peromyscus* mice, *Geomys* gophers, and other creatures) revealed three qualitative distributional patterns: (i) significant departure of lineage structure from equilibrium geographic distributions, a pattern exhibited in three rodent species with limited dispersal; (ii) nonsignificant departure from equilibrium expectations, exhibited by two avian and two marine fish species with potentials for relatively long-distance dispersal; and (iii) a progression from nonequilibrium distributions for younger lineages to equilibrium distributions for older lineages, a condition displayed by one surveyed avian species. These results demonstrate the advantages of considering mutation and genealogy when interpreting geographic variation in mtDNA.

14

Other Mammals

INTRODUCTION

Relatively few nonrodent mammals have been studied in JCA's laboratory, but this paucity is compensated by the fascinating nature of the other mammals that have been examined. These include several *Macaca* monkeys, which are renowned for their capacity to hybridize, and *Dasypus* armadillos that are unique among all mammals in their proclivity to produce genetically identical (polyembryonic) litters constitutively. As we shall see, both of these evolutionary phenomena lend themselves especially well to analysis via polymorphic molecular markers.

Allelic expression and genetic distance in hybrid macaque monkeys

Avise, J.C. and S.W. Duvall. 1977. Allelic expression and genetic distance in hybrid macaque monkeys. **Journal of Heredity** *68:23–30.*

ANECDOTE OR BACKDROP

By the mid-1970s, an exciting molecular notion had begun to gain traction: that changes in patterns of gene regulation (rather than structural changes in protein-coding genes) might typically be at the heart of adaptive evolution. There was one problem, however, with this alluring hypothesis: nobody knew quite how to identify or analyze regulatory genes directly. Thus, researchers began to suggest various indirect or surrogate approaches to the problem. For example, some authors proposed that the magnitudes of allelic repression or gene silencing in interspecific hybrids might indicate the degree to which genomes of the two parent species had diverged from one another in their regulatory apparatuses. This article on hybrid macaque monkeys was an early attempt to put this interesting hypothesis to the test. The blood samples used in this study came from captive animals housed at Emory University's Yerkes Primate Center north of Atlanta Georgia.

ABSTRACT

Levels of structural genic divergence at 21 loci encoding blood proteins were quantified in six macaque (*Macaca*) species, using standard techniques of starch gel electrophoresis. Genetic distances between all pairs of species fall within a narrow range ($0.08 < D < 0.25$;

mean $D = 0.16$) that is near the lower limit of genetic distances typically observed between other congeneric organisms. In an effort to measure levels of regulatory gene differences between these species, we have examined the patterns of allelic expression in their F_1, F_2, and backcross hybrids. Nine of the 21 loci examined encode allelic forms of the proteins with different electrophoretic mobilities in at least some of the individual parents of the hybrids. In all cases where expected, hybrids fully express both maternal and paternal allelic products, thus providing no strong evidence for a breakdown in the regulatory mechanisms responsible for proper expression of these genes. Results are compared to degrees of allelic repression previously observed in other hybrids (such as in fishes) and are discussed within the context of current ideas about rates of regulatory gene evolution in mammals.

ADDENDUM

JCA's colleague and primatologist Sue Duvall tragically died shortly after this project was completed, and just before the authors were ready to conduct a follow-up study that they had planned on lemurs.

FIGURE 14.1 Rhesus monkey, *Macaca mulatta*.

Biochemical variation and genetic heterogeneity in South Carolina deer populations

Ramsey, P.R., J.C. Avise, M.H. Smith, and D.F. Urbston. 1979. Biochemical variation and genetic heterogeneity in South Carolina deer populations. Journal of Wildlife Management 43:136–142.

ANECDOTE OR BACKDROP

The white-tailed deer is probably the most abundant large mammal in North America, and it poses hazards to people because of its proclivity to frequent roadsides and cause collisions with cars. To ameliorate this problem, and to thin overcrowded herds, officials at the Savannah River Plant (SRP) in South Carolina routinely staged formal deer hunts in which many scores of the animals were systematically shot by sportsmen. Being employees of the Savannah River Ecology Lab (SREL) at that time, Paul Ramsey and JCA were assigned the unpleasant task of gutting the slaughtered deer and gathering their internal organs for genetic analyses. That is how we came about obtaining such large sample sizes of white-tailed deer for the following protein-electrophoretic survey.

ABSTRACT

Protein variation in 2189 white-tailed deer (*Odocoileus virginianus*) from the Savannah River Plant in South Carolina was examined by starch gel electrophoresis. Polymorphism occurred for 7 of 21 structural loci coding for 20 protein systems and for a gene duplication of alpha-chain hemoglobin. Segregating alleles were detected for esterase, transferrin, phosphoglucomutase, glutamate oxalacetic transaminase, malate dehydrogenase, and beta-chain hemoglobin. Swamp and upland subpopulations were recognized based on 6 years of data from controlled hunts. The swamp herd had higher but declining density, an older age structure, 35% lower fertility among female fawns, and 13% greater mortality of male fawns at the time the genetic data were collected. Esterase and hemoglobin loci showed significant differences in genotypic proportions between herds, sexes, and age classes. Associated demographic and genetic differences suggest several applications of electrophoretic data to management practices by identifying subpopulations, assessing migration, and detecting selection.

ADDENDUM

In the western United States, the white-tailed deer is replaced by its close relative the mule deer (Odocoileus hemionus), which also has been the subject of extensive genetic surveys (by other labs). In North America, white-tailed deer are probably much more abundant now than they have ever been on the continent. The clearing of eastern forests during the 1800s and early 1900s opened much favorable edge habitat for this species.

FIGURE 14.2   White-tailed deer, *Odocoileus virginianus*.

Molecular documentation of polyembryony and the micro-spatial dispersion of clonal sibships in the nine-banded armadillo, *Dasypus novemcinctus*

Prodöhl, P.A., W.J. Loughry, C.M. McDonough, W.S. Nelson, and J.C. Avise. 1996. Molecular documentation of polyembryony and the micro-spatial dispersion of clonal sibships in the nine-banded armadillo, Dasypus novemcinctus. *Proceedings of the Royal Society of London B 263:1643–1649.*

ANECDOTE OR BACKDROP

The nine-banded Armadillo attracted our attention because it was thought to be essentially unique among all vertebrates in its consistent display of a particular form of clonality known as polyembryony (or monozygotic "twinning"). The phenomenon of constitutive polyembryony in a mammal had long puzzled evolutionary biologists because it appears at face value to combine unfavorable aspects of sexuality and asexuality. Each armadillo litter was long believed to consist of four genetically identical pups that nonetheless differ genetically from both parents. It seemed as if Mother Nature was pursuing a rather stupid tactic of making carbon copies of a multilocus nuclear genotype that had never before been field tested (like buying multiple copies of one lottery ticket). What was going on? Are armadillos truly polyembryonic, and if so, how might this peculiar type of clonality among full siblings have evolved? We hoped that molecular markers combined with field observations would help to answer such questions about this strange little mammal.

ABSTRACT

A battery of allelic markers at highly polymorphic microsatellite loci was developed and employed to genetically confirm the clonal nature of sibships in nine-banded armadillos. This phenomenon of consistent polyembryony, otherwise nearly unknown among the vertebrates, then was capitalized upon to describe the microspatial distributions of numerous clonal sibships in a natural population of armadillos. Adult clonemates were significantly more dispersed than were juvenile sibs, suggesting limited opportunities for altruistic behavioral interactions among mature individuals. These results, and considerations of armadillo natural history, suggest that evolutionary explanations for polyembryony in this species may not reside in the kinds of ecological and kin selection theories relevant to some of the polyembryonic invertebrates. Rather, polyembryony in armadillos may be associated evolutionarily with other reproductive peculiarities of the species, including delayed uterine implantation of a single egg.

ADDENDUM

Polyembryony is an intragenerational form of clonality, and thus stands in contradistinction to intergenerational forms of clonality such as parthenogenesis and its related evolutionary modes (see Chapter 3).

FIGURE 14.3   Nine-banded Armadillo, *Dasypus novemcinctus*.

Genetic maternity and paternity in a local population of armadillos assessed by microsatellite DNA markers and field data

Prodöhl, P.A., W.J. Loughry, C.M. McDonough, W.S. Nelson, E.A. Thompson, and J.C. Avise. 1998. Genetic maternity and paternity in a local population of armadillos assessed by microsatellite DNA markers and field data. **American Naturalist** *151:7—19.*

ANECDOTE OR BACKDROP

JCA's genetic work on armadillos was done in collaboration with Bill Loughry and Colleen McDonough, two researchers from Valdosta State University who had devoted much of their careers to field studies of armadillos in the southeastern United States. To conduct the genetic assays, JCA recruited to his lab a technologically savvy postdoc—Paulo Prodöhl—who proved to be a perfect choice for the genetic analyses involving newly discovered microsatellite loci. Paulo ran the most beautiful gels that we had ever seen!

ABSTRACT

Genetic data from polymorphic microsatellite loci were employed to estimate paternity and maternity in a local population of nine-banded armadillos (*Dasypus novemcinctus*) in northern Florida. The parentage assessments took advantage of maximum likelihood procedures developed expressly for situations when individuals of neither gender can be excluded *a priori* as candidate parents. The molecular data for 290 individuals, interpreted alone and in conjunction with detailed biological and spatial information for the population, demonstrate: (i) high exclusion probabilities and reasonably strong likelihoods of genetic parentage assignment in many cases; (ii) low mean probabilities of successful reproductive contribution to the local population by individual armadillo adults in a given year; and (iii) statistically significant microspatial associations of parents and their offspring. Results suggest that molecular assays of highly polymorphic genetic systems can add considerable power to assessments of biological parentage in natural populations, even when neither parent otherwise is known.

Correlates of reproductive success in a population of nine-banded armadillos

Loughry, W.J., P.A. Prödohl, C.M. McDonough, W.S. Nelson, and J.C. Avise. 1998. Correlates of reproductive success in a population of nine-banded armadillos. **Canadian Journal of Zoology 76:1815–1821.**

ANECDOTE OR BACKDROP

This was another in our series of genetic analyses, coupled with field observations and other evidence, on polyembryonic armadillos.

ABSTRACT

We used microsatellite DNA markers to identify the putative parents of 69 litters of nine-banded armadillos (*Dasypus novemcinctus*) over 4 years, and used that genetic parentage data as backdrop for the following morphological appraisals. Male and female parents did not differ in any measure of body size in comparisons with nonparents. However, males observed paired with a female were significantly larger than unpaired males, although paired females were the same size as unpaired females. Females categorized as possibly lactating were significantly larger than females that were either definitely lactating or definitely not lactating. There was no evidence of assortative mating: body size measurements of mothers were not significantly correlated with those of fathers. Nine-banded armadillos give birth to litters of genetically identical quadruplets. Mothers (but not fathers) of female litters were significantly larger than mothers of male litters, and maternal (but not paternal) body size was positively correlated with the number of surviving young within years, but not cumulatively. There were no differences in dates of birth between male and female litters, nor were there any significant relationships between birth date and maternal body size. Body size of either parent was not correlated with the body sizes of their offspring. Cumulative and yearly reproductive success did not differ between reproductively successful males and females. The majority of adults in the population apparently failed to produce any surviving offspring, and even those that did usually did so in only one of the 4 years. This low reproductive success is unexpected, given the rapid range expansion of this species throughout the southeastern United States in this century.

Polyembryony in armadillos

Loughry, W.J., P.A. Prodöhl, C.M. McDonough, and J.C. Avise. 1998. Polyembryony in armadillos. American Scientist 86:274–279.

ANECDOTE OR BACKDROP

This review article provided a summary of our multifaceted research program on the nine-banded Armadillo.

ABSTRACT

Recent advances in biotechnology raise the possibility of producing clonal offspring in a variety of vertebrates, and there has been considerable discussion about the ramifications of animal cloning. One way of assessing the consequences of such artificial manipulations is to examine species that make clones naturally. The behavioral and other data we have collected suggest that the production of clonal offspring by nine-banded armadillos, *Dasypus novemcinctus*, has had little impact on their behavior or ecology. Our data also cast doubt on simplistic models of kin selection that use genetic relatedness as the primary predictor of the incidence of altruism. Similar findings have been reported in other species. For instance, individuals of a parthenogenetic lizard, *Lepidodactylus lugubris*, are more aggressive toward one another than are individuals in a related species that does not produce clones. Such findings do not invalidate kin selection theory, but they do point out that the ecological and evolutionary situations can be more complicated than originally supposed.

ADDENDUM

For several years, JCA tried unsuccessfully to get tissue samples from several other species of Dasypus *armadillos native to South America. These other* Dasypus *species are likewise thought to be polyembryonic, but often have different litter sizes. This group of animals would still make a superb model system for studying the evolution and phylogeny of constitutive intragenerational clonality in a vertebrate taxon.*

Hemiplasy and homoplasy in the karyotypic phylogenies of mammals

Robinson, T.J., A. Ruiz-Herrera, and J.C. Avise. 2008. Hemiplasy and homoplasy in the karyotypic phylogenies of mammals. Proceedings of the National Academy of Sciences USA 105:14477–14481.

ANECDOTE OR BACKDROP

Hemiplasy was a new term that JCA invented to refer to situations in which a gene tree is discordant with a species tree due to idiosyncratic lineage sorting across successive speciation events. This paper—with South African colleague Terry Robinson—provides empirical examples of the phenomenon involving African mammals.

ABSTRACT

Phylogenetic reconstructions are often plagued by difficulties in distinguishing phylogenetic signal (due to shared ancestry) from phylogenetic noise or homoplasy (due to character-state convergences or reversals). We utilize a new interpretive hypothesis, termed hemiplasy, to show how random lineage sorting might account for specific instances of seeming "phylogenetic discordance" among different chromosomal traits, or between karyotypic features and probable species phylogenies. We posit that hemiplasy is generally less likely for underdominant chromosomal polymorphisms (i.e., those with heterozygous disadvantage) than for neutral polymorphisms or especially for overdominant rearrangements (which should tend to be longer lived), and we illustrate this concept using examples from chiropterans and afrotherians. Chromosomal states are especially powerful in phylogenetic reconstructions because they offer strong signatures of common ancestry, but their evolutionary interpretations remain fully subject to the principles of cladistics and the potential complications of hemiplasy.

ADDENDUM

In recent years, numerous examples have come to light in which the distributions of cytoplasmic genomes (mitochondrial DNA in animals, or chloroplast DNA in plants) do not precisely match the taxonomic boundaries of the creatures in which they are currently housed. Such phylogenetic discordances traditionally have been ascribed mostly to historical introgression following hybridization between particular pairs of species. However, an alternative explanation for such genetic discordances posits that evolutionary nodes in a phylogeny of the focal species were spaced closely together in time (relative to effective population size), such that particular gene-tree lineages could have sorted in such a way as to eventuate in the observed discord with current taxonomic boundaries. In other words, a gene tree could be discordant with a species tree for plausible evolutionary reasons other than historical hybridization alone. The term hemiplasy was introduced to describe this hypothetical phylogenetic phenomenon. It remains our hope that the word will someday become a popular shorthand for an evolutionary phenomenon that is potentially extremely important but underappreciated. This article provided one of the first well-documented empirical examples of hemiplasy in a real-life situation.

FIGURE 14.4   Two Afrotherian mammals: elephant and hyrax.

15

Invertebrates

INTRODUCTION

Although JCA is somewhat of a vertebrate chauvinist, his laboratory occasionally has delved into the invertebrate realm as well. The abstracts in this chapter reflect that research, most of which has involved various invertebrate animals from both aquatic (freshwater) and maritime (coastal marine) environments. Arranged chronologically, these papers form a progression that closely matches the succession of molecular tools available for providing genetic markers: protein methodologies and histocompatibility bioassays in the early years; restriction site and sequencing assays of mitochondrial DNA in the middle years; and microsatellite analyses of unlinked nuclear loci in more recent times. The topics that were addressed of course show a corresponding progression, from assessments of local (and sometimes clonal) population structures, to phylogeographic surveys across species' ranges, to species' phylogenies, to refined assessments of genetic paternity, maternity, and mating systems in brood-carrying taxa.

Enzyme changes during development of holo- and hemi-metabolic insects

Avise, J.C. and J.F. McDonald. 1976. Enzyme changes during development of holo- and hemi-metabolic insects. Comparative Biochemistry and Physiology 53B:393–397.

ANECDOTE OR BACKDROP

This study and the next were "C.V. padders," much as small Christmas gifts are sometimes referred to as "stocking stuffers." They were conducted while JCA and his close friend John McDonald were graduate students at U.C. Davis. Neither study had anything to do with our respective dissertation projects, but we felt it might be helpful to learn some new biochemical techniques (including spectrophotometric assays), and also to add some ancillary studies to our curriculum vitae for when the time came to graduate and apply for faculty positions.

ABSTRACT

Several carbohydrate-metabolizing enzymes were examined spectrophotometrically and electrophoretically in four life stages of the holometabolic fruit fly, *Drosophila pseudoobscura*, and in three life stages of the hemimetabolic pea aphid, *Acrythosiphon pisum*. *Drosophila* pupae exhibited significantly lower enzyme activities than third instar larvae. Most enzymes recovered to larval levels or higher in adults, but some continued to decrease following eclosion. Enzyme activities remained relatively unchanged during development of *Acrythosiphon*. The same isozymic forms of each enzyme were usually represented in all life stages of *Drosophila*, and in all life stages of *Acrythosiphon*. Enzyme levels paralleled the distinct morphological and physiological changes characterizing development of representative holo- and hemi-metabolic insects.

ADDENDUM

John McDonald also went on to a highly successful career in academic evolutionary genetics.

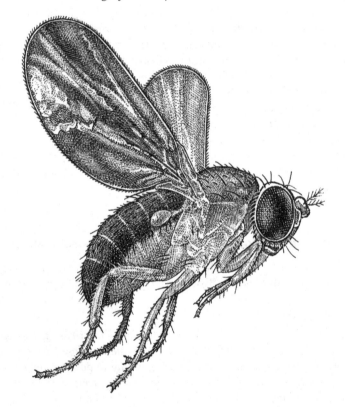

FIGURE 15.1   Fruit fly, *Drosophila pseudoobscura*.

Evidence for the adaptive significance of enzyme activity levels: interspecific variation in alpha-GPDH and ADH in *Drosophila*

McDonald, J.F. and J.C. Avise. 1976. Evidence for the adaptive significance of enzyme activity levels: interspecific variation in alpha-GPDH and ADH in Drosophila. **Biochemical Genetics 14:347–355.**

ABSTRACT

The activity levels of alcohol dehydrogenase and alpha-glycerophosphate dehydrogenase were compared among nine species of *Drosophila* representing three phylogenetic groups. For any given life stage, interspecific variability in activity level was much greater for ADH than for GPDH. Patterns of ontogenetic expression of enzyme activity were also much more variable among species for ADH than for GPDH. These results are consistent with the interpretation that GPDH is involved with a relatively uniform adaptive function among species, whereas ADH levels may reflect variable adaptive capabilities. There is a significant correlation between ADH activities and survivorship on alcohol-treated media for these nine species.

ADDENDUM

For a short time in the 1970s, several protein-electrophoretic laboratories focused on ADH in Drosophila *because this was deemed to be a favorable system for detecting natural selection mediated by an environmental agent (in this case the alcohol that arises from fermentation processes).*

Genetic variation in cave-dwelling and deep-sea organisms, with emphasis on *Crangonyx antennatus* (Crustacea: Amphipoda) in Virginia

Dickson, G.W., J.C. Patton, J.R. Holsinger, and J.C. Avise. 1979. Genetic variation in cave-dwelling and deep-sea organisms, with emphasis on Crangonyx antennatus *(Crustacea: Amphipoda) in Virginia.* **Brimleyana 2:119–130.**

ANECDOTE OR BACKDROP

This small study in effect provided a brief conceptual and empirical follow-up to JCA's Masters thesis on cave-dwelling fishes of the genus Astyanax (see Chapter 4*).*

ABSTRACT

Genetic variation was analyzed through electrophoretic techniques in six populations of the troglobitic (i.e., obligatory cave-dwelling) amphipod *Crangonyx antennatus* from Lee County, VA. From the results of this investigation and those tabulated from previous studies on a number of cave-dwelling species, genetic variability does not appear to be substantially reduced in populations inhabiting subterranean environments. The origin of normal levels of genetic variation in cave-dwelling species may differ from those of organisms inhabiting another relatively stable environment: the deep sea. The high levels of genetic variability recorded in many deep-sea invertebrates are thought to be due in part to the presence of large populations of these species. In contrast, the small population sizes observed in cave-dwelling organisms may allow species to expand their niches with an associated increase in genetic variability.

ADDENDUM

I no longer believe that caves or the deep sea provide quintessential examples of stable and uniform environments.

FIGURE 15.2   Cave Amphipod, *Crangonyx antennatus.*

Clonal diversity and population structure in a reef-building coral, *Acropora cervicornis*: self-recognition analysis and demographic interpretation

Neigel, J.E. and J.C. Avise. 1983. Clonal diversity and population structure in a reef-building coral, Acropora cervicornis: *self-recognition analysis and demographic interpretation.* Evolution 37:437–453.

ANECDOTE OR BACKDROP

This paper, which involves the use of tissue grafts to identify clones, is representative of several such articles on corals and sponges that JCA coauthored with Joe Neigel, one of his earliest and most productive graduate students. This research grew out of our initial frustration in attempts to find polymorphic allozyme markers in these marine invertebrates that can reproduce both sexually and asexually (clonally). Lacking suitable molecular markers to distinguish clones, we resorted to the use of histocompatibility bioassays, an approach that proved to be a blessing in disguise. Not only did such in situ bioassays prove to be remarkably powerful for our goal of distinguishing clones and mapping their spatial arrangements, but they also provided some of the earliest (and still most convincing) evidence for the radical notion that various invertebrates possess self-recognition genetic systems that might even rival the precision of the immunological and histocompatibility systems of vertebrate animals.

ABSTRACT

Biological self-recognition phenomena, analogous in many respects to vertebrate histocompatibility, are apparently widespread in scleractinian corals. Here we exploit the operational properties of a self-recognition bioassay to estimate clonal diversity and population structure in a common reef-building coral, *Acropora cervicornis*. When branches of *A. cervicornis* come into contact, within several months the tissues and calcareous matrix exhibit either an acceptance or rejection response. Through a variety of controlled experimental grafts, and observations of natural interactions, we demonstrate that these responses accurately distinguish clonemates (the products of asexual fragmentation of colonies) from nonclonemates (the products of sexual recruitment). This self-recognition bioassay was subsequently used to analyze clonal structure and diversity in *A. cervicornis* populations in Discovery Bay, Jamaica, and Tague Bay, St. Croix, U.S. Virgin Islands. A total of nearly 500 experimental grafts were scored *in situ* and additional information was obtained from observations of naturally occurring contacts. In comparison to the Tague Bay population, the Discovery Bay population has a significantly greater microgeographical clonal diversity that is reflected in a much higher Neighbor Index (the proportion of rejection responses in contacts among adjacent colonies). In Tague Bay, spatial maps of *A. cervicornis* clones were constructed; they show that clones are variable in size (from a single colony to assemblages up to 10 meters or more in diameter), and are distributed in discrete patches. A computer model was used to simulate the development of clonal population structure from simple demographic processes (colony mortality, asexual and sexual recruitment). Using empirically derived rate estimates of these demographic processes we found that the quantitative predictions of the Neighbor Index values from the model corresponded closely to our bioassay estimates of this parameter for real populations. Qualitative predictions of the model were also in good agreement with our empirical assessments of the size range of an individual clone's spatial distribution in the Discovery Bay population, and the degree of segregation and variance in size observed for spatial distributions mapped in the Tague Bay population. The model indicated that the relatively low diversity of clones in the Tague Bay population is expected when the input of new clones into the population by sexual recruitment is extremely low. Recently settled *Acropora* colonies did in fact appear to be virtually absent in this population. Sedimentation on the Tague Bay reef may interfere with the establishment of juvenile scleractinia on the available substrata. The simulations also predict that occasions of catastrophic mortality and subsequent recovery of *A. cervicornis* populations may be major determinants of clonal diversity and population structure, especially when the rate of sexual recruitment is relatively low. Major disturbances, such as severe hurricanes, may prevent populations of *A. cervicornis* from attaining equilibrium patterns of clonal structure.

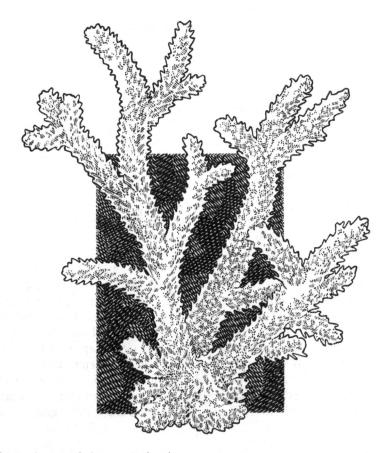

FIGURE 15.3 Staghorn coral, *Acropora cervicornis*.

Histocompatibility bioassays of population structure in marine sponges: clonal structure in *Verongia longissima* and *Iotrochota birotulata*

Neigel, J.E. and J.C. Avise. 1983. Histocompatibility bioassays of population structure in marine sponges: clonal structure in Verongia longissima *and* Iotrochota birotulata. Journal of Heredity 74:134–140.

ANECDOTE OR BACKDROP

All of our histocompatibility work on corals and sponges were conducted in situ on reefs in Jamaica or St. Croix. Over periods typically lasting several weeks, our usual daily regimen was to SCUBA dive for 2–3 hours every morning and afternoon, laboriously cutting off and then grafting (tying with monofilament line attached to a plastic tag) colonial branches to other colonies in a wide variety of experimental designs. We did many hundreds of grafts in this fashion, on several branching species of soft sponges and hard corals. We would then return to the same reef site weeks or months later (depending on the coral or sponge species) to score the grafting results. This intensive regimen of SCUBA diving unquestionably provided the most enjoyable research of JCA's entire career. And the scientific findings themselves were equally exciting!

ABSTRACT

Clonal population structure in two marine sponges, *Verongia longissima* and *Iotrochota birotulata*, was examined with a self-recognition bioassay. The bioassay consists of grafts of branch segments between conspecific individuals. Results were consistent with the operational properties expected of a precise histocompatibility system: autografts exhibited acceptance responses, grafts between individuals separated by large distances exhibited rejection responses, individuals were not limited to a single mode of response at one time, and all identity relationships were transitive. Clonal population structure was assessed by examining the relationship between response and donor-to-recipient distance, and by actually mapping the distributions of particular clones. Clones of *Iotrochota birotulata* were usually restricted to single coral heads or small patch reefs (1–3 meters in diameter). For *Verongia longissima*, which can grow directly upon the coral rubble surrounding coral heads and patch reefs, individual clones often occupied larger areas (up to 10 meters diameter). The spatial patterns of clonal distributions are readily interpreted as consequences of the particular demographies and habitat specificities of these two species.

ADDENDUM

In our experience, the histocompatibility responses in sponges were logistically superior to those in corals because they matured (and hence could be scored) in a matter of weeks rather than months. This offered obvious advantages with respect to our design of grafting experiments on the reef.

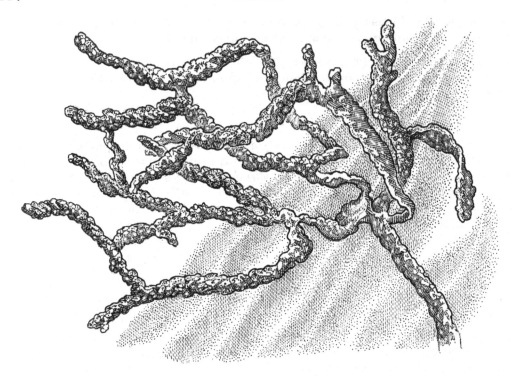

FIGURE 15.4   Green Finger Sponge, *Iotrochota birotulata*.

The precision of histocompatibility response in clonal recognition in tropical marine sponges

Neigel, J.E. and J.C. Avise. 1985. The precision of histocompatibility response in clonal recognition in tropical marine sponges. Evolution 39:724–732.

ANECDOTE OR BACKDROP

This was another in our series of publications on histocompatibility responses and clonal recognition in tropical marine corals and sponges.

ABSTRACT

Recently discovered histocompatibility-like phenomena in sponges (phylum Porifera) have prompted attempts to measure the precision with which allogeneic grafts are recognized and rejected. The results of these investigations have been extremely varied, ranging from suggestions that allorecognition does not occur to suggestions that every genetically distinct individual may be unique in histocompatibility type. Interpretation of these findings is complicated by the variation in methods and species used by different workers. Here we compare various measurements of allorecognition precision for several species of tropical marine sponges. From our results we conclude that: (i) tissue implant grafts are more prone to artifact than grafts between intact sponges; (ii) the possibility of clonal propagation should be considered when graft acceptances are observed between sponges taken from a single population; and (iii) allozyme variation in *Niphates erecta* shows that grafts between genetically different individuals occasionally may be accepted.

Population biology aspects of histocompatibility polymorphisms in marine invertebrates

Avise, J.C. and J.E. Neigel. 1984. Population biology aspects of histocompatibility polymorphisms in marine invertebrates. Pp. 223–234 in: **Genetics: New Frontiers, V.L. Copra, B.C. Joshi, R.P. Sharma, and H.C. Banoal (eds.). Oxford Press, New Delhi.**

ANECDOTE OR BACKDROP

This article summarized our own (and others) multiyear experiences with histocompatibility responses in various species of corals and sponges.

ABSTRACT

The literature on recognition of foreignness in invertebrates is large but of obviously variable quality. Different researchers have worked with different species, different life-history stages, or have employed very different experimental protocols. Several studies have documented extreme specificity in allogenic responses, while other reports continue to be negative. In almost no instances have studies been replicated. Some observations in the literature have an anecdotal quality. While it may yet prove true that nearly every species or group will be highly idiosyncratic in capacity to respond to foreign tissue contacts, it seems doubtful to us that all of the apparent heterogeneity in histocompatibility response will ultimately be attributable to real differences in biological self-recognition capabilities. In our work with several species of tropical reef corals and sponges, we have been most impressed with the similarities in response across species, and with the nearly invariate concordance of grafting results with the operational properties expected for a precise self-recognition system. Until recently, the population units identified in most histocompatibility studies were referred to simply as "strains" (exceptions involve the work of Hildemann and colleagues). This designation was conservative, and also carried little motivation for further conceptual or experimental work. The clonal nature of histo-compatibility strains will certainly have to be further tested, utilizing either grafting experiments in lineages with known biological pedigrees, or by the use of a sufficient number of independent genetic markers for clonal assay. However, in view of the considerable evidence that now exists about the pattern and specificity of response for many tropical marine sponges and cnidarians, the working hypothesis that histocompatibility assays distinguish clones is justified at this time. Even if the hypothesis proves not to be entirely correct, it should provide a powerful heuristic construct for more rapid advance in both the molecular and population aspects of the field.

ADDENDUM

In recent decades, researchers have worked out the precise genetic and mechanistic bases of histo-compatibility responses in various invertebrate taxa.

Critical experimental test of the possibility of "paternal leakage" of mitochondrial DNA

Lansman, R.A., J.C. Avise, and M.D. Huettel. 1983. Critical experimental test of the possibility of "paternal leakage" of mitochondrial DNA. **Proceedings of the National Academy of Sciences USA** *80:1969–1971.*

ANECDOTE OR BACKDROP

Mitochondria typically display maternal inheritance in sexually reproducing species, but questions persisted about the possibility and ramifications of occasional paternal leakage of mtDNA into matrilines via sperm. We were well aware of the importance of such issues even in the early years of our mtDNA research, and it occurred to us that a powerful way to assess mtDNA inheritance empirically would be to analyze progeny from experimental lines produced by multiple generations of unidirectional backcrossing with respect to gender (see Chapter 3*). This paper further explains the rationale behind this simple idea and presents the results of one such critical test for paternal leakage—in tobacco budworms! We had exhaustively searched the literature for any species for which such unidirectional backcrossing had been conducted for many successive generations, so we were delighted to find this peculiar example in a small moth species.*

ABSTRACT

Most previous data suggesting maternal inheritance of mtDNA have come from single generation mating experiments, and most of the analytical techniques utilized would not have detected paternal mtDNA molecules in progeny at levels less than about 5%. Long-term mating experiments, in which a fertile female lineage derived from hybridization of two species with distinguishable mtDNAs is backcrossed recurrently to the male parental species, provide an ideal opportunity to assess possible low-level paternal leakage. We have analyzed the 45-generation and 91-genration backcross progeny of such matings between two species of lepidopteran insects [*Heliothis* (Noctuidae)], using auto-radiographic techniques that can detect rare mtDNA molecules in less than 1 part per 500. The analysis failed to detect any paternal mtDNA and sets an upper limit of paternal leakage at about 1 molecule per 25,000 per generation in this system.

ADDENDUM

The insect strains used in this study came from long-term experiments conducted by the U.S. Department of Agriculture in Gainesville, FL., as part of their efforts to find a means to eradicate this serious pest species.

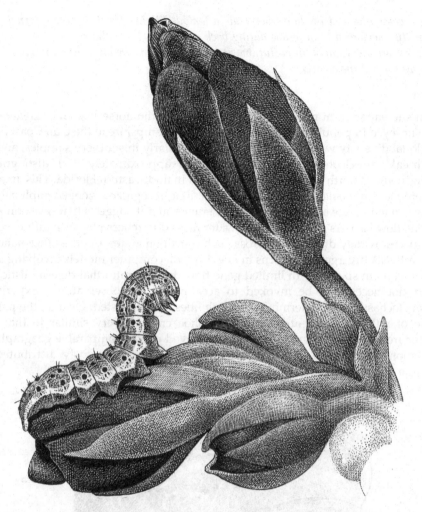

FIGURE 15.5   Tobacco budworm, *Heliothis virescens*.

Genetic variation and geographic differentiation in mitochondrial DNA of the horseshoe crab, *Limulus polyphemus*

Saunders, N.C., L.G. Kessler, and J.C. Avise. 1986. Genetic variation and geographic differentiation in mitochondrial DNA of the horseshoe crab, Limulus polyphemus. *Genetics 112:613–627.*

ANECDOTE OR BACKDROP

In the early years of the mtDNA revolution in population genetics, most of our genetic surveys involved vertebrates (such as various mammals and fishes), but we soon wondered whether mtDNA assays and phylogeographic perspectives could likewise be applied to invertebrate animals. One of our first test cases involved the horseshoe crab, a famous "living fossil" whose external phenotype seems virtually unchanged from fossils dating back more than 150 million years. Did such extreme phenotypic conservatism imply an evolutionary conservatism in the animals' genotypes as well? In this paper, we learned the answer.

ABSTRACT

Restriction site variation in mitochondrial (mt) DNA of the horseshoe crab (*Limulus polyphemus*) was surveyed in populations ranging from New Hampshire to the Gulf Coast of Florida. mtDNA clonal diversity was moderately high, particularly in southern samples, and a major genetic "break" (nucleotide sequence divergence approximately 2%) distinguished all sampled individuals north versus south of a region in northeastern Florida. This area of genotypic divergence in *Limulus* corresponds to a long-recognized zoogeographic boundary between warm temperate and tropical marine faunas, and it suggests that selection pressures and/or gene flow barriers associated with water mass differences may also influence the evolution of species widely distributed across such transition zones. On the other hand, a comparison of mtDNA divergence patterns in *Limulus* with computer models involving stochastic lineage extinction in species with limited gene flow demonstrates that deterministic explanations need not necessarily be invoked to account for the observations. Experiments to distinguish stochastic from deterministic possibilities are suggested. Overall, the pattern and magnitude of mtDNA differentiation in horseshoe crabs is very similar to that typically reported for freshwater and terrestrial species assayed over a comparable geographic range. Results demonstrate for the first time that at least some continuously distributed marine organisms can show considerable geographic mtDNA genetic differentiation.

FIGURE 15.6   Horseshoe crab, *Limulus polyphemus*.

A speciational history of "living fossils": molecular evolutionary patterns in horseshoe crabs

Avise, J.C., W.S. Nelson, and H. Sugita. 1994. A speciational history of "living fossils": molecular evolutionary patterns in horseshoe crabs. Evolution 48:1986–2001.

ANECDOTE OR BACKDROP

The world has four living species of horseshoe crabs—one in the Americas and three from southeast Asia. One day when we opened a large package that had been shipped to us from Japan, we were amazed to see first hand how morphologically similar the Asian species were to our own familiar American crabs. Except for very minor differences such as cross-sectional profiles of the tail or telson, all of these four species looked nearly identical to us. They also closely resembled now-extinct creatures that roamed the seas more than 150 million years ago. This paper was the second of our genetic reports on these amazing representatives of "living fossils."

ABSTRACT

Horseshoe crabs' exceptional morphological conservatism over the past 150 million years has led to their reputation as "living fossils," but also has served to obscure phylogenetic relationships within the complex. Here we employ nucleotide sequences from two mitochondrial genes to assess molecular evolutionary rates and patterns among all extant horseshoe crab species. The American species *Limulus polyphemus* proved to be the sister taxon to a clade composed of the Asiatic species *Tachypleus gigas*, *T. tridentatus*, and *Carcinoscorpius rotundicauda*, whose relationships *inter se* were not resolved definitively. Both absolute and relative rate tests suggest a moderate slowdown in sequence evolution in horseshoe crabs. Nonetheless, dates of the lineage separations remain uncertain primarily because of reservations about molecular clock calibrations resulting from large rate variances at examined loci across Arthropods and other animal lineages, as inferred in this and prior studies. Thus, ironically, separation dates as estimated by molecular evidence may remain most insecure in taxonomic groups for which such information is most sorely needed—those lacking strong biogeographic or fossil benchmarks for internal clock calibrations. In any event, the current results show that large numbers of molecular characters distinguish even these most morphologically conservative of organisms. Furthermore, comparisons against previously published mitochondrial sequence data in the morphologically dynamic hermit crab plus king crab taxonomic complex demonstrate that striking heterogeneity in levels of morphotypic differentiation can characterize Arthropod lineages at similar magnitudes of molecular divergence.

Phylogenetic assessment of length variation at a microsatellite locus

Ortí, G., D.E. Pearse, and J.C. Avise. 1997. Phylogenetic assessment of length variation at a microsatellite locus. Proceedings of the National Academy of Sciences USA 94:10745–10749.

ANECDOTE OR BACKDROP

This was a third paper that emerged from our early genetic appraisals of horseshoe crabs. Postdoc Guillermo Ortí led the way on this one, which unlike the others were focused more squarely on molecular rather than organismal-level phenomena.

ABSTRACT

Sixty-six haplotypes at a locus containing a simple dinucleotide $(CA)_n$ microsatellite repeat were isolated by PCR single-strand conformational polymorphism from populations of the horseshoe crab *Limulus polyphemus*. These haplotypes were sequenced to assess nucleotide variation directly. Thirty-four distinct sequences (alleles) were identified in a region 570 bp long that included the microsatellite motif. In the repeat region itself, CA number varied in integer values from 5 to 11 across alleles, except that a $(CA)_8$ class was not observed. Differences among alleles were due also to polymorphisms at 22 sites in regions immediately flanking the microsatellite repeats. Nucleotide substitutions in these regions were used to estimate phylogenetic relationships among alleles, and the gene phylogeny was used to trace the evolution of length variation and CA repeat numbers. A low correlation between size variation and genealogical relationships among alleles suggests that absolute fragment size (as normally scored in microsatellite assays) is an unreliable indicator of historical affinities among alleles. This finding on the molecular fine structure of microsatellite variation suggests the need for caution in the use of repeat counts at microsatellite loci as secure indicators of allelic relationships.

A genetic discontinuity in a continuously distributed species: mitochondrial DNA in the American oyster, *Crassostrea virginica*

Reeb, C.A. and J.C. Avise. 1990. A genetic discontinuity in a continuously distributed species: mitochondrial DNA in the American oyster, Crassostrea virginica. *Genetics 124:397–406.*

ANECDOTE OR BACKDROP

Oysters are abundant and easy to collect along coastal shorelines in the southeastern United States. Thus, when Carol Reeb joined JCA's lab during the mid-1980s, she immediately set to work on examining mitochondrial genotypes in this species, the intent basically being twofold: (i) to further extend our explorations of mtDNA in the invertebrate realm and (ii) to compare any phylogeographic results to those we had previously published for various maritime vertebrates in the same region of the country. Carol amply succeeded on both counts.

ABSTRACT

Restriction site variation in mitochondrial (mt) DNA of the American oyster (*Crassostrea virginica*) was surveyed in continuously distributed populations sampled from the Gulf of St. Lawrence, Canada to Brownsville, TX. mtDNA clonal diversity was high, with 82 different haplotypes revealed among 212 oysters with 13 endonucleases. The mtDNA clones grouped into two distinct genetic arrays (estimated to differ by about 2.6% in nucleotide sequence) that characterized oysters collected north versus south of a region on the

Atlantic mid-coast of Florida. The population-genetic "break" in mtDNA contrasts with previous reports of near uniformity of nuclear (allozyme) allele frequencies throughout the range of the species, but agrees closely with the magnitude and pattern of mtDNA differentiation reported in other estuarine species in the southeastern United States. This concordance of mtDNA phylogeographic pattern across independently evolving species provides strong evidence for vicariant biogeographic processes initiating intraspecific population structure. The post-Miocene ecological history of the region suggests that reduced precipitation levels in an enlarged Floridian peninsula may have created disconti-nuities in suitable estuarine habitat for oysters during glacial periods, and that today such population separations are maintained by the combined influence of ecological gradients and oceanic currents on larval dispersal. The results are consistent with the hypothesis that historical vicariant events, in conjunction with contemporary environmental influ-ences on gene flow, can result in genetic discontinuities in continuously distributed species with high dispersal capability.

FIGURE 15.7   American oyster, *Crassostrea virginica*.

Balancing selection at allozyme loci in oysters: implications from nuclear RFLPs

Karl, S.A. and J.C. Avise. 1992. Balancing selection at allozyme loci in oysters: implications from nuclear RFLPs. Science 256:100–102.

ANECDOTE OR BACKDROP

I've already mentioned (see Chapter 9) that graduate student Steve Karl was becoming an expert on some newer RFLP (restriction fragment length polymorphism) methodologies for uncovering molecular variation from the nuclear genome. Here he again showcases these skills by contrasting RFLP findings in the American oyster against those previously reported for allozymes.

ABSTRACT

Population genetic analyses that depend on the assumption of neutrality for allozyme markers are used widely. Restriction fragment length polymorphisms in nuclear DNA of the American oyster evidence a pronounced population subdivision concordant with mitochondrial DNA. This finding contrasts with a geographic uniformity in allozyme frequencies previously thought to reflect high gene flow mediated by the pelagic gametes and larvae. The discordance likely is due to selection on protein-electrophoretic characters that balances allozyme frequencies in the face of severe constraints to gene flow. These results raise a cautionary note for studies that rely on assumptions of neutrality for allozyme markers.

PCR-based assays of Mendelian polymorphisms from anonymous single-copy nuclear DNA: techniques and applications for population genetics

Karl, S.A. and J.C. Avise. 1993. PCR-based assays of Mendelian polymorphisms from anonymous single-copy nuclear DNA: techniques and applications for population genetics. **Molecular Biology and Evolution 10:342–361.**

ANECDOTE OR BACKDROP

This paper formalizes and details some of the newer laboratory techniques that Steve Karl was developing, as a part of his dissertation, for revealing nuclear genetic polymorphisms.

ABSTRACT

This paper outlines a PCR-based approach for population genetics that offers several advantages over conventional Southern blotting methods for revealing restriction fragment length polymorphisms (RFLPs) in nuclear DNA. Primers are constructed from clones isolated from a nuclear DNA library, and these primers subsequently are employed in *in vitro* syntheses of homologous regions. Amplified products are then screened directly for RFLPs by using gel-staining procedures. Population applications for this PCR-based approach, including potential strengths and weaknesses, are exemplified by two RFLP data sets generated to estimate: (i) male-mediated gene flow in the green turtle (*Chelonia mydas*) and (ii) geographic population genetic structure in the American oyster (*Crassostrea virginica*). Restriction assays of amplified products from 14 or 15 independent primer pairs in each species revealed polymorphisms at several loci that proved highly informative in the population genetic analyses. In general, the Mendelian polymorphisms produced by this PCR-based approach will provide useful genetic markers for population studies, particularly in situations where simpler and less expensive allozyme methods have failed, for whatever reason, to provide adequate information.

ADDENDUM

In recent years, the laboratory methods described in this paper have largely been supplanted by new sequencing technologies.

Population structure in the American oyster as inferred by nuclear gene genealogies

Hare, M.P. and J.C. Avise. 1998. Population structure in the American oyster as inferred by nuclear gene genealogies. **Molecular Biology and Evolution** *15:119–128.*

ANECDOTE OR BACKDROP

Matt Hare was another of my outstanding graduate students in the 1990s, and for several years "the oyster was his world." Matt's assignment was to develop and employ cytonuclear markers to study a transitional zone between genetically distinct oyster populations in the Atlantic Ocean versus the Gulf of Mexico. This paper and the ones that follow describe some of the fruits of Matt's labor.

ABSTRACT

Multiple haplotypes from each of three nuclear loci were isolated and sequenced from geographic populations of the American oyster, *Crassostrea virginica*. In tests of alternative phylogeographic hypotheses for this species, nuclear gene genealogies constructed for these haplotypes were compared to one another, to a mitochondrial gene tree, and to patterns of allele frequency variation in nuclear restriction site polymorphisms (RFLPs) and allozymes. Oyster populations from the Atlantic versus the Gulf of Mexico are not reciprocally monophyletic in any of the nuclear gene trees, despite considerable genetic variation and despite large allele frequency differences previously reported in several other genetic assays. If these populations were separated vicariantly in the past, either insufficient time has elapsed for neutral lineage sorting to have achieved monophyly at most nuclear loci, or balancing selection may have inhibited allelic extinction, or secondary gene flow may have moved haplotypes between regions. These and other possibilities are examined in light of available genetic evidence, and it is concluded that no simple explanation can account for the great variety of population genetic patterns across loci displayed by American oysters. Regardless of the source of this heterogeneity, this study provides an empirical demonstration that different DNA sequences within the same organismal pedigree can have quite different phylogeographic histories.

ADDENDUM

It remains true that nuclear gene genealogies are rather seldom reported in phylogeographic surveys, and that such data are difficult to obtain.

Molecular genetic analysis of a stepped multilocus cline in the American oyster (*Crassostrea virginica*)

Hare, M.P. and J.C. Avise. 1996. Molecular genetic analysis of a stepped multilocus cline in the American oyster (Crassostrea virginica). **Evolution** *50:2305–2315.*

ANECDOTE OR BACKDROP

This study further extended our genetic surveys of maritime invertebrates in the southeastern United States.

ABSTRACT

Gulf of Mexico versus Atlantic populations of several coastal species in the southeastern United States are known to differ sharply in genetic composition, but most transitional zones have not previously been examined in detail. Here we employ molecular markers from mitochondrial and nuclear loci to characterize cytonuclear genetic associations at meso- and microgeographic scales along an eastern Florida transitional zone between genetically distinct Atlantic and Gulf populations of the American oyster, *Crassostrea virginica*. The single- and multilocus cytonuclear patterns display: (i) a cline extending along 340 km of the east Florida coastline; (ii) a pronounced step in the cline (shifts in allele frequencies by 50–75% over a 20-km distance centered at Cape Canaveral); (iii) a close agreement of observed genotypic frequencies with Hardy–Weinberg expectations within locales; and (iv) mild or nonexistent nuclear and cytonuclear disequilibria in most local population samples. These results imply: (i) considerable restrictions to interpopulation gene flow along the eastern Florida coastline; (ii) within locales, free interbreeding (as opposed to mere population admixture) between Gulf and Atlantic forms of oysters; and (iii) localized population recruitment in the transition zone localities. These findings demonstrate that marine organisms with high dispersal potential via long-lived pelagic larvae can nonetheless display pronounced spatial population genetic structure, and more generally they exemplify the utility of pronounced genetic transition zones for the study of population-level processes.

Anonymous nuclear DNA markers in the American oyster and their implications for the heterozygote deficiency phenomenon in marine bivalves

Hare, M.P., S.A. Karl, and J.C. Avise. 1996. Anonymous nuclear DNA markers in the American oyster and their implications for the heterozygote deficiency phenomenon in marine bivalves. **Molecular Biology and Evolution 13:334–345.**

ANECDOTE OR BACKDROP

This was one of our follow-up studies on the American oyster, focusing on a peculiar population genetic phenomenon previously reported in some marine bivalves.

ABSTRACT

A puzzling population-genetic phenomenon widely reported in allozyme surveys of marine bivalves is the occurrence of heterozygote deficits relative to Hardy–Weinberg expectations. Possible explanations for this pattern are categorized with respect to whether the effects should be confined to protein-level assays or are genomically pervasive and expected to be registered in both protein-level and DNA-level assays. Anonymous nuclear DNA markers from the American oyster were employed to reexamine the phenomenon. In assays based on the polymerase chain reaction (PCR), two DNA-level processes were encountered that can lead to artifactual genotypic scorings: (i) differential amplification of alleles at a target locus and (ii) amplification from multiple paralogous loci. We describe symptoms of these complications and prescribe methods that should generally help to ameliorate them. When artifactual scorings at two anonymous DNA loci in the American oyster were corrected, Hardy–Weinberg deviations registered in preliminary population assays decreased to nonsignificant values. Implications of these findings for the heterozygote-deficit phenomenon in marine bivalves, and for the general development and use of PCR-based assays, are discussed.

Genetic parentage assessment in the crayfish *Orconectes placidus*, a high-fecundity invertebrate with extended maternal brood care

Walker, D., B.A. Porter, and J.C. Avise. 2002. Genetic parentage assessment in the crayfish Orconectes placidus, *a high-fecundity invertebrate with extended maternal brood care. Molecular Ecology 11:2115–2122.*

ANECDOTE OR BACKDROP

As repeatedly emphasized in Chapters 2–6, *genetic parentage analyses are greatly facilitated when one of the two parents of a brood is known or suspected from independent (nonmolecular) evidence. Such is the case in female-pregnant mammals or fishes, for example, because each embryo's mother is then obvious and the father's genotype can then be deduced "by subtraction" using suitable genetic markers. Many other animals likewise have analogues of female pregnancy, and the same logic then applies with respect to deducing paternity. This paper exemplifies my laboratory's efforts to expand our microsatellite-based parentage analyses beyond various vertebrates and into the invertebrate realm.*

ABSTRACT

Microsatellite data recently have been introduced in the context of genetic maternity and paternity assignments in high-fecundity fish species with single parent-tended broods. Here we extend such analyses to an aquatic invertebrate, the crayfish *Orconectes placidus*, in which gravid females carry large numbers of offspring. Genetic parentage analyses of more than 900 progeny from 15 wild crayfish broods revealed that gravid females were invariably the exclusive dams of the offspring they tended (i.e., there was no allomaternal care), and that most of the females had mated with multiple (usually two) males who contributed sometimes highly skewed numbers of offspring to a brood. Within any multiple sired brood, the unhatched eggs (or the hatched juveniles) from different fathers were randomly distributed across the mother's brood space. All of these genetic findings are discussed in the light of observations on the mating behaviors and reproductive biology of crayfishes.

ADDENDUM

These animals were obtained simply as a byproduct of our collecting efforts for several nest-tending stream fishes in the southeastern United States (see Chapters 2–4).

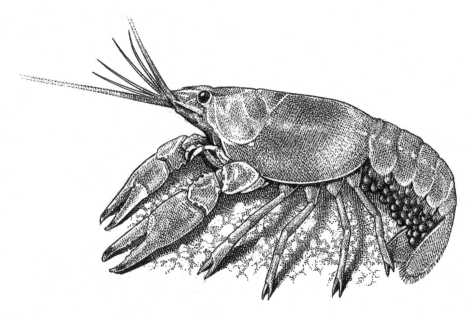

FIGURE 15.8   Bigclaw crayfish, *Orconectes placidus*.

Genetic sex determination, gender identification, and pseudohermaphroditism in the knobbed whelk, *Busycon carica* (Mollusca; Melongenidae)

Avise, J.C., A.J. Power, and D. Walker. 2004. Genetic sex determination, gender identification, and pseudohermaphroditism in the knobbed whelk, Busycon carica *(Mollusca; Melongenidae).* **Proceedings of the Royal Society of London B 271:641–646.**

ANECDOTE OR BACKDROP

The Knobbed Whelk is a gastropod mollusk often seen on intertidal mudflats of the southeastern United States. Indeed, that is precisely the environment from which we collected these familiar animals along Georgia shorelines. Wading up to our knees through the mud, we simply picked up gravid females who we caught in the process of laying their egg-case strings, composed of a series of leathery capsules each containing as many as dozens of developing embryos. Thus, the mother of each string is known and one approachable genetic task is to deduce who may have sired particular broods.

ABSTRACT

We report perhaps the first genic-level molecular documentation of a mammalian-like "X-linked" mode of sex determination in molluscs. From family inheritance data and observed associations between sex-phenotyped adults and genotypes in *Busycon carica*, we deduce that a polymorphic microsatellite locus (*bc2.2*) is diploid and usually heterozygous in females, hemizygous in males, and that its alleles are transmitted from mothers to sons and daughters but from fathers to daughters only. We also employ *bc2.2* to estimate near-conception sex ratio in whelk embryos, where gender is undeterminable by visual inspection. Statistical corrections are suggested at both family and population levels to accommodate the presence of homozygous *bc2.2* females that could otherwise be genetically mistaken for hemizygous males. Knobbed whelks were thought to be sequential hermaphrodites, but our evidence for genetic dioecy supports an earlier hypothesis that whelks are pseudohermaphroditic (falsely appear to switch functional sex when environmental conditions induce changes in sexual phenotype). These findings highlight the distinction between gender in a genetic sense versus a phenotypic sense.

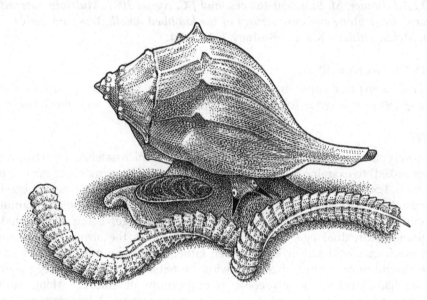

FIGURE 15.9   Knobbed whelk, *Busycon carica*.

Sex-linked markers facilitate genetic parentage analyses in knobbed whelk broods

Walker, D., A.J. Power, and J.C. Avise. 2005. Sex-linked markers facilitate genetic parentage analyses in knobbed whelk broods. Journal of Heredity 92:108–113.

ANECDOTE OR BACKDROP

This was our second paper on genetic parentage and mating systems in knobbed whelk broods.

ABSTRACT

To explore the potential of sex-linked polymorphisms for genetic parentage analyses in natural populations, we have employed a recently discovered "X-linked" microsatellite marker (in conjunction with polymorphic autosomal loci) to deduce biological paternity and maternity for large numbers of encapsulated embryos within individual broods of the knobbed whelk, *Busycon carica*. Empirical findings illustrate how such sex-linked genetic tags can in special instances find at least three novel utilities in genetic dissections of large-clutch species: clarification of paternity assignments that had remained ambiguous from dilocus autosomal data alone; elucidation of linkage relationships among pairs of autosomal loci; and illumination of maternity (and thereby paternity also) in broods for which neither biological parent was known from independent evidence.

Multiple paternity and female sperm usage along egg-case strings of the knobbed whelk, *Busycon carica* (Mollusca; Melongenidae)

Walker, D., A.J. Power, M. Sweeney-Reeves, and J.C. Avise. 2007. Multiple paternity and female sperm usage along egg-case strings of the knobbed whelk, Busycon carica *(Mollusca; Melongenidae).* **Marine Biology** *151:53–61.*

ANECDOTE OR BACKDROP

This was our third and final paper on genetic parentage and mating systems in knobbed whelks, an invertebrate species that proved to be utterly fascinating for its peculiar reproductive systems.

ABSTRACT

We used genotypic data from three highly polymorphic microsatellite loci (two autosomal and one sex-linked) to examine microspatial and temporal arrangements of genetic paternity for more than 1,500 embryos housed along 12 egg-case strings of the knobbed whelk, *Busycon carica*. Multiple paternity proved to be the norm in these single dam families, with genetic contributions of several sires (at least 3.5 on average) being represented among embryos within individual egg capsules as well as along the string. Two strings were studied in much greater detail; five and seven fathers were identified, none of which was among the several males found in consort with the female at her time of egg-laying. Each deduced sire had fathered roughly constant proportions of embryos along most of the string, but those proportions differed consistently among fathers. A few significant paternity shifts at specifiable positions along an egg-case string were also observed. Although the precise physical mechanisms inside a female whelk's reproductive tract remain unknown, our genetic findings indicate that successive fertilization events (and/or depositions of zygotes into egg capsules) normally occur as near-random draws from a well-but-not-perfectly blended pool of gametes (or zygotes) stemming from stored ejaculates (perhaps in different titers) of a dam's several mates.

ADDENDUM

Many other melongenid whelks similarly lay egg-case strings, and likewise would be good subjects for genetic appraisals of biological parentage.

Polygynandry and sexual size dimorphism in the sea spider *Ammothea hilgendorfi* (Pycnogonida: Ammotheidae), a marine arthropod with brood-carrying males

Barreto, F.S. and J.C. Avise. 2008. Polygynandry and sexual size dimorphism in the sea spider Ammothea hilgendorfi *(Pycnogonida: Ammotheidae), a marine arthropod with brood-carrying males.* **Molecular Ecology** *17:4164–4175.*

ANECDOTE OR BACKDROP

In 2005, JCA moved from Georgia to his current position at the University of California at Irvine. His first graduate student—Felipe Barreto—at this new institution decided to conduct his dissertation research on species that were abundant and native to southern California. Felipe's organisms of choice were sea spiders, a ubiquitous but poorly known group of marine Arthropods in which males carry clusters of fertilized eggs on their legs and thereby in effect can be thought of as displaying the otherwise rare phenomenon of "male pregnancy" (not too unlike the situation for pipefishes and seahorses as described in Chapter 5*). The final three abstracts in this chapter describe Felipe's findings with regard to genetic parentage, mating systems, and sexual selection in these tiny but fascinating intertidal and subtidal sea spiders. Felipe collected the pregnant animals simply by turning over small boulders or rocks at extreme low tides.*

ABSTRACT

Species that exhibit uniparental postzygotic investment by males are potentially good systems for investigating the interplay of sexual selection, parental care, and mating systems. In all species of sea spiders (Class Pycnogonida), males exclusively provide postzygotic care by carrying fertilized eggs until they hatch. However, the mating systems of sea spiders are poorly known. Here we describe the genetic mating system of a sea spider (*Ammothea hilgendorfi*) by assaying nearly 1,400 embryos from a total of 13 egg-carrying males across 4 microsatellite markers. We also determine the extent of sexual dimorphism in trunk and leg size, and assess how reproductive success in males varies with these morphological traits. We detected instances of multiple mating by both sexes, indicating that this species has a polygynandrous mating system. Genotypic assays also showed that: males do not mix eggs from different females in the same clusters; eggs from the same female are often partitioned into several clusters along a male's oviger; and clusters are laid chronologically from proximal to distal along ovigers. Females were on average larger than males with respect to leg length and width and trunk length, whereas males had wider trunks. Among the genotyped egg-carrying males, neither the number of eggs carried nor the number of mates was correlated with body size traits. Nevertheless, the high variance in mating success, genetically documented, suggests that males differ in their ability to acquire mates, so future studies are needed to determine what traits are the targets of sexual selection in this species. In addition to providing the first description of the mating system in a sea spider, our study illustrates the potential uses of this group for testing hypotheses from parental investment and sexual selection theories.

ADDENDUM

Felipe raised his sea spiders in the lab prior to the DNA preparations. Each fertilized egg and early embryo of a sea spider is very tiny (barely visible to the naked eye). Nevertheless, it carries ample DNA for successful amplification via the PCR (polymerase chain reaction).

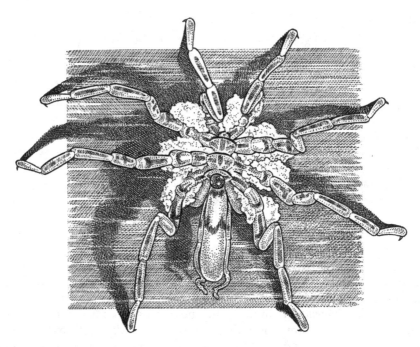

FIGURE 15.10   Sea spider, *Ammothea hilgendorfi*.

Quantitative measures of sexual selection reveal no evidence for sex-role reversal in a sea spider with prolonged paternal care

Barreto, F.S. and J.C. Avise. 2010. Quantitative measures of sexual selection reveal no evidence for sex-role reversal in a sea spider with prolonged paternal care. **Proceedings of the Royal Society of London B 277:2951–2956.**

ABSTRACT

Taxa in which males alone invest in postzygotic care of offspring are often considered good models for investigating the proffered relationships between sexual selection and mating systems. In the pycnogonid sea spider *Pycnogonum stearnsi*, males carry large egg masses on their bodies for several weeks, so this species is a plausible candidate for sex-role reversal (greater intensity of sexual selection on females than on males). Here we couple a microsatellite-based assessment of the mating system in a natural population with formal quantitative measures of genetic fitness to investigate the direction of sexual selection in *P. stearnsi*. Both sexes proved to be highly polygamous and showed similar standardized variances in reproductive and mating successes. Moreover, the fertility (number of progeny) of males and females appeared to be equally and highly dependent on mate access, as shown by similar Bateman gradients for the two sexes. The absence of sex-role reversal in this population of *P. stearnsi* is probably attributable to the fact that males are not limited by brooding space but have evolved an ability to carry large numbers of progeny. Body length was not a good predictor of male mating or reproductive success, so future studies should aim at determining what traits are the targets of sexual selection in this species.

The genetic mating system of a sea spider with male-biased sexual size dimorphism: evidence for paternity skew despite random mating success

Barreto, F.S. and J.C. Avise. 2011. The genetic mating system of a sea spider with male-biased sexual size dimorphism: evidence for paternity skew despite random mating success. Behavioral Ecology and Sociobiology *65:1595–1604.*

ABSTRACT

Male-biased size dimorphism is usually expected to evolve in taxa with intense male–male competition for mates, and it is hence associated with high variances in male mating success. This association, however, is thought to vary with many ecological and mating system characteristics, such as parental investment, adult sex ratios, spatial and temporal distribution of mates, and potential reproductive rates. Most species of pycnogonid sea spiders exhibit female-biased size dimorphism and are notable among arthropods for having exclusive male parental care of embryos. Here we first show that *Ammothella biunguiculata*, a small intertidal sea spider, is male-biased size dimorphic with respect to trunk dimensions. Moreover, we combine genetic parentage analysis with quantitative measures of sexual selection to show that male body size does not appear to be under directional selection. Comparison to null models of random mating revealed that mate acquisition in this species is largely driven by chance factors, although actual fertilization success is non-randomly distributed. In addition, the opportunity for sexual selection in *A. biunguiculata* was less than half of that estimated in a sea spider with female-biased size dimorphism, suggesting the direction of size dimorphism may not be a reliable predictor of the intensity of sexual selection in this group.

ADDENDUM

There exist more than 1000 described living species of sea spiders (Pycnogonidae), so many opportunities remain for further genetic analyses of these fascinating animals.

Intergroup Comparisons and Theory

INTRODUCTION

Astute readers will have noticed that nearly all of the scientific papers from JCA's laboratory involve explicit comparisons of evolutionary genetic or population genetic patterns in different taxa, such as evolutionary rates in cyprinid minnows versus centrarchid sunfishes (see Chapters 2 and 4), genetic mating systems in various pipefishes versus seahorse species (see Chapter 5), gene flow regimes in mouthbrooding catfishes versus demersal-spawning toadfishes (see Chapter 6), genetic variation in freshwater versus marine versus anadromous fishes (see Chapter 6), phylogeographic structures in various marine and freshwater turtle species (see Chapters 9 and 10), historical influences on populations of desert chuckwallas versus desert tortoises (see Chapter 11), genetic divergence in mimic thrushes versus vireos (see Chapter 12), and histocompatibility responses in branching corals versus branching sponges (see Chapter 15). Sometimes the empirical or theoretical comparisons involve even more distant taxa—such as between the various organismal groups that form the chapter headings for this book. This concluding chapter encapsulates several studies in which we explicitly compared various evolutionary genetic patterns observed or perhaps expected in creatures as different as peromyscine rodents versus parulid birds, pregnant fishes versus pregnant mammals, and viviparous vertebrates versus invertebrate brooders.

Systematic value of electrophoretic data

Avise, J.C. 1974. Systematic value of electrophoretic data. Systematic Zoology 23:465–481.

ANECDOTE OR BACKDROP

This overview article was published while JCA was still a graduate student at U.C. Davis. It was motivated by his (and the field's) growing realization that allozyme data could be very useful for taxonomic and phylogenetic purposes. This widely cited article initially was rejected, but in retrospect the author is glad that he persevered in seeing it through to eventual publication.

ABSTRACT

Two consistent observations from recent comparative multilocus electrophoretic studies are: (i) levels of genic similarity between conspecific populations appear very high (populations nearly identical in allelic content at 85 percent or more of their loci) and (ii) genic similarities between different, even very closely related species, are generally much lower and more widely dispersed (congeneric species pairs often completely distinct at one-fifth to four-fifths of their loci). These observations have valuable implications regarding the practical utility of electrophoresis: (i) one or a few samples often yield adequate data for the description of a species for systematic purposes and (ii) closely related species may be arranged according to percentages of shared alleles or genotypes. A survey of the literature indicates that when such arrangements are made, they usually correspond very closely to previously recognized relationships of various species groups based on classical systematic criteria. This observation, coupled with several theoretical advantages of the study of allozymes, makes it clear that protein-electrophoretic techniques will provide an extremely valuable tool for systematists.

Genetic change and rates of cladogenesis

Avise, J.C. and F.J. Ayala. 1975. Genetic change and rates of cladogenesis. **Genetics** *81:757–773.*

ANECDOTE OR BACKDROP

This is the paper in which the authors generated formal mathematical models and developed the conceptual context for critically distinguishing between phyletic gradualism and punctuated equilibrium based on genetic comparisons of extant members of rapidly versus slowly speciating clades (see Chapters 2 and 4).

ABSTRACT

Models are introduced that predict ratios of mean levels of genetic divergence in species-rich versus species-poor phylads under two competing assumptions: (i) genetic differentiation is a function of time, unrelated to the number of cladogenetic events and (ii) genetic differentiation is proportional to the number of speciation events in the group. The models are simple, general, and biologically real, but not precise. They lead to qualitatively distinct predictions about levels of genetic divergence depending upon the relationship between rates of speciation and amount of genetic change. When genetic distance between species is a function of time, mean genetic distances in speciose and depauperate phylads of equal age are very similar. On the contrary, when genetic distance is a function of the number of speciations in the history of a phylad, the ratio of mean genetic distances separating species in speciose versus depauperate phylads is greater than 1 and increases rapidly as the frequency of speciations in one group relative to the other increases. The models may be tested with data from natural populations to assess: (i) possible correlations between rates of anagenesis and cladogenesis and (ii) the amount of genetic differentiation accompanying the speciation process. The data collected in electrophoretic surveys and other kinds of studies can be used to test predictions of the models. For this purpose, genetic distances need to be measured in speciose and depauperate phylads of equal evolutionary age. The limited information presently available agrees better with the model predicting that genetic change is primarily a function of time and is not correlated with rates of speciation. Further testing of the models is, however, required before firm conclusions can be drawn.

Evolutionary genetics of birds. III. Comparative molecular evolution in New World warblers (Parulidae) and rodents (Cricetinae)

Avise, J.C., J.C. Patton, and C.F. Aquadro. 1980. Evolutionary genetics of birds. III. Comparative molecular evolution in New World warblers (Parulidae) and rodents (Cricetinae). Journal of Heredity 71:303–310.

ANECDOTE OR BACKDROP

This paper made explicit comparisons in mean levels of genetic divergence among New World warblers versus those among New World rodents. It was the third in a long series of papers (see Chapter 12) documenting a widespread "conservative" pattern of protein evolution in birds vis-à-vis many other organismal groups.

ABSTRACT

The American wood warblers (Parulidae) exhibit considerable diversity in breeding plumage coloration, song, behavior, ecology, and general life history. Nonetheless a comparison of 28 species representing 12 genera discloses a very conservative pattern of protein differentiation as gauged by standard protein-electrophoretic procedures. We define conservative to mean simply that at equivalent levels of the taxonomic hierarchy, parulids exhibit far smaller genetic distances than do most other organisms surveyed. This observation is documented and dramatized by comparing results with those obtained in the same laboratory, using similar electrophoretic conditions, on a group of American rodents of even lower taxonomic rank—the Cricetinae. Our study represents an extension and elaboration of similar findings, reported earlier, on parulid warblers. One possible explanation for this conservative pattern is that warbler speciations have been very recent. However, estimated divergence times for Parulidae, read from protein clocks calibrated for nonavian vertebrates and invertebrates, are much lower than estimates derived from a prevailing view of parulid zoogeography and evolution. It appears likely that protein evolution is decelerated in the wood warblers. If protein clocks generally prove to exhibit organism-dependent calibration, their usefulness in determining sidereal divergence times for species with a poor fossil record will be compromised.

ADDENDUM

New World warblers (Parulidae) are a species-rich group of beautifully plumaged creatures, not to be confused with the equally speciose but unrelated Old World warblers (Sylviidae) that generally are a drably-plumaged lot.

FIGURE 16.1   Canada Warbler, *Wilsonia canadensis.*

A comparative summary of genetic distances in the vertebrates: patterns and correlations

Avise, J.C. and C.F. Aquadro. 1982. A comparative summary of genetic distances in the vertebrates: patterns and correlations. Evolutionary Biology *15:151–185.*

ANECDOTE OR BACKDROP

This paper summarized the emerging protein-electrophoretic literature on genetic distances among related species representing a wide variety of vertebrate taxa.

ABSTRACT

Considering the multivarious factors that might be expected to influence mean genetic distances (D) in different genera, it would be surprising if any consistent trends in magnitude of D per genus were apparent among the vertebrates. Nonetheless, our summary of the multilocus protein-electrophoretic literature, including over 3,800 pairwise comparisons of species, has disclosed some strong and consistent patterns of genetic differentiation. In particular, avian congeners are extremely conservative in magnitude of protein divergence relative to amphibian congeners assayed to date. Few reptiles have been assayed, but three of five genera exhibit a nonconservative divergence pattern approaching that of the amphibians. Fish and mammalian genera are highly variable in D, but generally fall intermediate in level of divergence to the birds and amphibians. Several possible explanations for these trends are considered: (i) differing taxonomic conventions in different vertebrate groups; (ii) different proteins assayed; (iii) heterogeneous protein clocks; and (iv) various other explanations. Whatever the reasons for the heterogeneity in D's among the vertebrates, the results are of significance to interpretations of taxonomic assignment. If the protein-electrophoretic results are corroborated by studies of DNA, it will be abundantly clear that an average genus of amphibians, for example, is by no means equivalent to an average genus of birds from a genetic point of view. In fact, particular amphibian genera probably encompass a genetic diversity characteristic of avian families. We conclude by providing an example of how these and related considerations importantly influence views of comparative organismal relationship. Other authors have demonstrated that humans and chimpanzees are very similar genetically, with an average human polypeptide more than 99% identical to its chimpanzee counterpart. From a "frog's point of view" this is indeed a very small genetic distance, and it has raised the question of whether humans and chimpanzees warrant their current placement in different taxonomic families. However, from a bird's perspective the human−chimpanzee distance ($D = 0.62$) is quite substantial—larger than the distance between any pair of avian congeners assayed to date. Systematic decisions should of course be based on many kinds of information, including morphology, physiology, and so forth. But because molecular data often provide quantitative, common scales for comparing disparate organisms, they should play an important role in our perceptions of the evolutionary relationships of organisms summarized in existing taxonomies.

ADDENDUM

The genetic trends reported in this paper are ripe for updating and reappraisal, based on the far more extensive comparative sequence data that are now available for many genes.

Models of mitochondrial DNA transmission genetics and evolution in higher eukaryotes

Chapman, R.W., J.C. Stephens, R.A. Lansman, and J.C. Avise. 1982. Models of mitochondrial DNA transmission genetics and evolution in higher eukaryotes. **Genetical Research** *40:41−57.*

ANECDOTE OR BACKDROP

This was one of our early attempts to model the intra- and intercellular dynamics of populations of mitochondrial DNA molecules in eukaryotic cells. This theoretical effort was led by Robert Chapman, then a graduate student in JCA's laboratory.

ABSTRACT

The future value of mitochondrial DNA sequence information to studies in population biology will depend in part on an understanding of mtDNA transmission genetics both within cell lineages and between animal generations. A series of stochastic models has been constructed here based on various possibilities concerning this transmission. Several of the models generate predictions inconsistent with available data and, hence, their assumptions are provisionally rejected. Other models cannot yet be falsified. These latter models include assumptions that: (i) mtDNAs are sorted through cellular lineages by random allocation to daughter cells in germ cell lineages, (ii) the effective intracellular population sizes (n_{Ms}) of mtDNA are small, and (iii) sperm may (or may not) provide a low level "gene flow" bridge between otherwise isolated female lineages. It is hoped that these models have helped to identify and will stimulate further empirical study of various parameters likely to strongly influence mtDNA evolution. In particular, critical experiments or measurements are needed to determine the effective sizes of mtDNA populations in germ (and somatic) cells and to examine possible parental contributions to zygote mtDNA composition.

ADDENDUM

Several key parameters (such as mtDNA population sizes in intermediate germ cell lineages) remain poorly known even to this day.

Polymorphism of mitochondrial DNA in populations of higher animals

Avise, J.C. and R.A. Lansman. 1983. Polymorphism of mitochondrial DNA in populations of higher animals. Pp. 147–164 in: Evolution of Genes and Proteins, M. Nei and R.K. Koehn (eds.). Sinauer, Sunderland, MA.

ANECDOTE OR BACKDROP

This was an early review of our developing laboratory experiences and empirical findings regarding mitochondrial DNA.

ABSTRACT

Several findings on mtDNA polymorphism within species of higher animals were unanticipated: the high level of polymorphism itself; the apparent rarity of individual heteroplasmy; the attribution of most polymorphism to base substitution; the preponderance of transitions; and the frequency of convergent site evolution. Thus, a general outline for the evolution of higher animal mtDNA is now apparent. A further recent surprise is the different scenario emerging for plant mtDNA. For example, maize mtDNA is approximately 30 times larger than most animal mtDNA and consists of at least 7 classes of molecules that vary in size and abundance. In maize and teosinte mtDNA, sequence homology is generally conserved and most polymorphism is attributable to major reorganizations of sequence. The evolutionary significance of the different pattern of plant mtDNA polymorphism is not known. Recent studies suggest that the level of mtDNA polymorphism in animals is several-fold higher than that of single copy nuclear DNA. Some of these conclusions resulted from indirect comparisons of protein-electrophoretic data with data from mtDNA restriction digests. Because restriction enzyme approaches can also be applied to measure polymorphism in the nuclear genome, direct answers to the question of relative levels of mtDNA versus nuclear DNA polymorphism should be forthcoming. Whatever the final outcome, mtDNA polymorphism will continue to be of special interest in its own right. The maternal inheritance and high polymorphism of mtDNA provide unique opportunities for population-level analyses.

Demographic influences on mitochondrial DNA lineage survivorship in animal populations

Avise, J.C., J.E. Neigel, and J. Arnold. 1984. Demographic influences on mitochondrial DNA lineage survivorship in animal populations. **Journal of Molecular Evolution** *20:99–105.*

ANECDOTE OR BACKDROP

This paper was among the first to highlight the special connection that exists between historical population demography of females and the genealogical relationships of mitochondrial DNA. Indeed, these two phenomena are like opposite sides of the same coin.

ABSTRACT

Probability models of branching processes and computer simulations of these models are used to examine stochastic survivorship of female lineages under a variety of demographic scenarios. A parameter π, defined as the probability of survival of two or more independent lineages over G generations, is monitored as a function of founding size of a population, population size at carrying capacity, and the frequency distributions of surviving progeny. Stochastic lineage extinction can be very rapid under certain biologically plausible demographic conditions. For stable-sized populations initiated by n females and/or regulated about carrying capacity $k = n$, it is highly probable that within about $4n$ generations all descendents will trace their ancestries to a single founder female. For a given mean family size, increased variance decreases lineage survivorship. In expanding populations, however, lineage extinction is dramatically slowed, and the final k value is a far more important determinant of π than is the size of the population at founding. Results are discussed in the context of recent empirical observations of low mitochondrial DNA sequence heterogeneity in humans and expected distributions of asexually transmitted traits among sexually reproducing species.

ADDENDUM

Coalescent theory is the thriving modern branch of evolutionary biology that deals explicitly with genetics—demography connections.

Phylogenetic relationships of mitochondrial DNA under various demographic models of speciation

Neigel, J.E. and J.C. Avise. 1986. Phylogenetic relationships of mitochondrial DNA under various demographic models of speciation. Pp. 515–534 in: Evolutionary Processes and Theory, *E. Nevo and S. Karlin (eds.). Academic Press, New York, NY.*

ANECDOTE OR BACKDROP

Although Joe Neigel worked primarily on histocompatibility responses and clonal population structure in marine corals and sponges (see Chapter 15*), he was also a gifted computer programmer and theoretician. These latter skills are on display in this paper, in which we showed that the phylogenetic status of mitochondrial lineages in sister taxa is itself evolutionarily dynamic. This conclusion was rather shocking for population geneticists at that time, but is now taken nearly for granted as a part of accepted evolutionary wisdom.*

ABSTRACT

Recent empirical studies have revealed instances in which maternally inherited mitochondrial (mt) DNAs of some populations are more closely related to those of a distinct biological species than to those of other conspecifics. One plausible explanation involves secondary hybridization and introgression. Here we examine whether phylogenetic sorting of matriarchal lineages, independent of any hybridization, can in principle also give rise to discordancies between biological species boundaries and mtDNA genotype. Computer models are developed to simulate the evolutionary dynamics of matriarchal lineages during formation of daughter species from an ancestral parent population. The relative probabilities of monophyly, paraphyly, and polyphyly of the daughter species (their "phylogenetic status" with respect to mtDNA) are monitored as a function of time since speciation under a variety of demographic scenarios. Major results are as follows: (i) phylogenetic distributions of mtDNA can lack concordance with species boundaries when species are recently separated; (ii) the phylogenetic status of a given pair of species is itself a dynamic evolutionary characteristic, with a common time-course of changes subsequent to speciation being polyphyly → paraphyly → monophyly; and (iii) the demographic mode of speciation has a major influence on the developing phylogenetic status of related species.

Mitochondrial DNA and the evolutionary genetics of higher animals

Avise, J.C. 1986. Mitochondrial DNA and the evolutionary genetics of higher animals. Philosophical Transactions of the Royal Society London B *312:325–342.*

ANECDOTE OR BACKDROP

This was another early review on the emerging notion that mitochondrial mtDNA could become a uniquely powerful molecular marker for evolutionary geneticists.

ABSTRACT

Mitochondrial (mt) DNA in higher animals is rapidly becoming a well-characterized system at the molecular level. Here I shift the focus to consider questions in organismal evolution that can be addressed by mtDNA assays. For the first time, it is possible to estimate matriarchal phylogeny empirically, to determine directionality in crosses producing hybrids, and to study the population genetic consequences of varying female demographies and life histories. The data obtainable from mtDNA may be especially well suited for studies of population genetic structure, dispersal, and historical zoogeography. The female-mediated, clonal transmission of mtDNA is also stimulating new ways of thinking about times to common ancestry of asexual lineages within otherwise sexually reproducing populations; about the possible relevance of mtDNA–nuclear DNA interactions to reproductive isolation; and about the very meaning of the phylogenetic status of related species with respect to particular kinds of genetic characters.

ADDENDUM

Today, mitochondrial DNA remains arguably the most powerful genetic system for a wide range of evolutionary applications.

Intraspecific phylogeography: the mitochondrial DNA bridge between population genetics and systematics

Avise, J.C., J. Arnold, R.M. Ball, Jr., E. Bermingham, T. Lamb, J.E. Neigel, C.A. Reeb, and N.C. Saunders. 1987. Intraspecific phylogeography: the mitochondrial DNA bridge between population genetics and systematics. **Annual Review of Ecology and Systematics** *18:489–522.*

ANECDOTE OR BACKDROP

This is the paper that laid the foundation for the field of "phylogeography." It coined the term, enunciated phylogeographic concepts and principles, and generally gave birth to this new discipline.

ABSTRACT

Mitochondrial DNA has provided the first extensive and readily accessible data available to evolutionists in a form suitable for strong genealogical inference at the intraspecific level. The rapid pace of mtDNA nucleotide substitution, coupled with the special mode of maternal nonrecombining inheritance, offers advantages for phylogenetic analysis at the microevolutionary level that will not be matched easily by any nuclear gene system. These peculiarities of mtDNA data have literally forced the addition of a phylogenetic perspective to studies of intraspecific evolutionary process and as such have provided an empirical and conceptual bridge between the nominally rather separate disciplines of systematics and population genetics. mtDNA has also served to clarify thought about the distinction between (yet relevance of) gene genealogies to organismal phylogeny. Many species have proved to exhibit a deep and geographically structured mtDNA phylogenetic history. Study of the relationship between genealogy and geography constitutes a discipline that can be termed intraspecific phylogeography. We present several phylogeographic hypotheses that were motivated by available data and that represent possible trends whose broader generality remains to be tested. Study of intraspecific phylogeography should assume a place in evolutionary biology at least commensurate with that of ecogeography, with mutual benefit resulting to both disciplines. Theories of speciation and macroevolution must now recognize and accommodate the reality of phylogeographic differentiation at the intraspecific level.

ADDENDUM

Phylogeography is now well entrenched as a fundamental scientific branch of biogeography and evolutionary genetics.

Gene trees and organismal histories: a phylogenetic approach to population biology

Avise, J.C. 1989. Gene trees and organismal histories: a phylogenetic approach to population biology. Evolution 43:1192—1208.

ANECDOTE OR BACKDROP

This was another early review of the keystone distinction between gene trees and population-level phylogenies.

ABSTRACT

A "gene tree" is the phylogeny of alleles or haplotypes for any specified stretch of DNA. Gene trees are components of population trees or species trees; their analysis entails a shift in perspective from many of the familiar models and concepts of population genetics, which typically deal with frequencies of phylogenetically unordered alleles. Molecular surveys of haplotype diversity in mitochondrial DNA have provided the first extensive empirical data suitable for estimation of gene trees on a microevolutionary (intraspecific) scale. The relationship between phylogeny and geographic distribution constitutes the phylogeographic pattern for any species. Observed phylogeographic trees can be interpreted in terms of historical demography by comparison to predictions derived from models of gene lineage sorting, such as inbreeding theory and branching-process theory. Results of such analyses for more than 20 vertebrate species strongly suggest that the demographies of populations have been remarkably dynamic and unsettled over space and recent evolutionary time. This conclusion is consistent with ecological observations documenting dramatic population size fluctuations and range shifts in many contemporary species. By adding a historical perspective to population biology, the gene lineage approach can help forge links between the disciplines of phylogenetic systematics (macroevolution) and population genetics (microevolution). Preliminary extensions of the "gene tree" methodology to haplotypes of nuclear genes (such as *Adh* in *Drosophila melanogaster*) demonstrate that the phylogenetic perspective can also help to illuminate molecular-genetic processes (such as recombination or gene conversion), as well as contribute to knowledge of the origin, age, and molecular basis of particular adaptations.

Principles of genealogical concordance in species concepts and biological taxonomy

Avise, J.C. and R.M. Ball, Jr. 1990. Principles of genealogical concordance in species concepts and biological taxonomy. Oxford Surveys in Evolutionary Biology 7:45–67.

ANECDOTE OR BACKDROP

Species concepts were hotly contested throughout the latter third of the twentieth century, with several alternative species conceptions being proffered by various authors. This paper represented JCA's attempt to wade into the debate.

ABSTRACT

One important root of the PSC (phylogenetic species concept) probably traces to George Gaylord Simpson's paleontological perspective on taxa, summarized in his definition of an evolutionary species as "a lineage (ancestral-descendant sequence of populations) evolving separately from others and with its own unitary roles and tendencies." Current versions of the PSC, apparently motivated by a perceived lack within the BSC (biological species concept) of an adequate emphasis on history and phylogeny, have led some PSC proponents to reject the BSC's emphasis on reproductive isolation. Principles of genealogical concordance provide a compromise or composite stance between the BSC and the PSC. Concepts of concordance are far from new in systematics, and numerous statements can be found in support of the desirability of concordant information prior to taxonomic recognition of a species. Yet such sentiments too seldom have been followed, and many taxa continue to be recognized on the basis of one or a few diagnostic traits. The new generation of systematists can avoid a repeat of such errors by requiring concordance among several independent characters before advocating formal taxonomic recognition of putative population disjunctions. By focusing on the phylogenetic consequences of intrinsic RBs (reproductive barriers), and by emphasizing that important historical partitions can also be present within a biological species because of extrinsic RBs, concordance principles should provide a useful set of philosophical and operational guidelines for the recognition of biotic species and taxonomic diversity.

ADDENDUM

Concordance principles have recently become quite widely adopted by systematists.

Gene genealogies within the organismal pedigrees of random-mating populations

Ball, R.M., Jr., J.E. Neigel, and J.C. Avise. 1990. Gene genealogies within the organismal pedigrees of random-mating populations. Evolution 44:360–370.

ANECDOTE OR BACKDROP

As its title suggests, this study explored the theory of a nonanastomose gene tree within a panmictic population.

ABSTRACT

Using computer simulations, we generated and analyzed genetic distances among selectively neutral haplotypes transmitted through gene genealogies within random-mating organismal pedigrees. Constraints and possible biases on haplotype distances due to correlated ancestry were evaluated by comparing observed distributions of distances to those predicted from an inbreeding theory that assumes independence among haplotype pairs. Results suggest that: (i) mean time to common ancestry of neutral haplotypes can be a reasonably good predictor of evolutionary effective population size; (ii) the nonindependence of haplotype paths of descent within a given gene genealogy typically produces significant departures from the theoretical probability distributions of haplotype distances; and (iii) frequency distributions of genetic distances between haplotypes drawn from "replicate" organismal pedigrees or from multiple unlinked loci within an organismal pedigree exhibit very close agreement with the theory for independent haplotypes. These results are relevant to interpretations of current molecular data on genetic distances among nonrecombining haplotypes at either nuclear or cytoplasmic loci.

Ten unorthodox perspectives on evolution prompted by comparative population genetic findings on mitochondrial DNA

Avise, J.C. 1991. Ten unorthodox perspectives on evolution prompted by comparative population genetic findings on mitochondrial DNA. **Annual Review of Genetics** *25:45−69.*

ANECDOTE OR BACKDROP

This was yet another early review article on the atypical evolutionary perspectives afforded by mitochondrial DNA.

ABSTRACT

Novel insights of general significance frequently stem from the study of atypical biological systems. Animal mtDNA is a classic example of an "aberrant" genetic system, whose predominant uniparental and nonrecombinational mode of transmission has stimulated evolutionary perspectives that differ from those typifying traditional Mendelian and population genetics. Several important concepts and research areas in evolutionary genetics—such as concern with organelle origins, the evolutionary strategies underlying intergenomic interactions, the notion of intraspecific phylogeography, and the distinction between gene trees and organismal phylogenies—were all motivated in large part by studies of animal mtDNA. In the future, such fresh perspectives might be applied profitably to a reexamination of nuclear genomes as well, for example, in considering the phylogenetic histories of particular nuclear loci, or the coevolutionary interactions between sex-linked chromosomes and autosomes. The tiny DNA sequences sequestered in a cell's mitochondria will probably continue to have a disproportionately large impact on evolutionary genetic thought.

Recognizing the forest for the trees: testing temporal patterns of cladogenesis using a null model of stochastic diversification

Wollenberg, K., J. Arnold, and J.C. Avise. 1996. Recognizing the forest for the trees: testing temporal patterns of cladogenesis using a null model of stochastic diversification. Molecular Biology and Evolution 13:833–849.

ANECDOTE OR BACKDROP

There was a rather long history in paleontology of using null models from computer simulations as a backdrop for interpreting empirical data sets on topics such as rates and patterns of speciation and extinction as registered in the fossil record. This paper extended such null evolutionary models in ways that could make them relevant and applicable to molecular data sets gathered from extant species. These computer models were then applied to molecular phylogenies estimated for creatures ranging from plants to insects and birds.

ABSTRACT

Computer simulations are developed and employed to examine the expected temporal distributions of nodes under a null model of stochastic lineage bifurcation and extinction. These Markovian models of phylogenetic process were constructed so as to permit direct comparisons against empirical phylogenetic trees generated from molecular or other information available solely from extant species. For replicate simulated phylads with n extant species, cumulative distribution functions (CDFs) of branching times were calculated and compared (using the Kolmogorov–Smirnov test statistic D) to those from three published empirical trees. Molecular phylogenies for columbine plants and avian cranes showed statistically significant departures from the null expectations, in directions indicating recent and ancient species' radiations, respectively, whereas a molecular phylogeny for the *Drosophila virilis* species group showed no apparent historical clustering of branching events. Effects of outgroup choice and phylogenetic frame of reference were investigated for the columbines and found to have a predictable influence on the types of conclusions to be drawn from such analyses. To statistically test for nonrandomness in temporal cladogenetic pattern in empirical trees generated from data on extant species, we present tables of mean CDFs and associated probabilities under the null model for expected branching times in phylads of varying size. These approaches complement and extend other recent methods for employing null models to assess the statistical significance of branching patterns in evolutionary trees.

Phylogenetics and the origin of species

Avise, J.C. and K. Wollenberg. 1997. Phylogenetics and the origin of species. Proceedings of the National Academy of Sciences USA 94:7748–7755.

ANECDOTE OR BACKDROP

Darwin's original treatise "On the Origin of Species" was followed nearly a century later with a genetic update by Theodosius Dobzhansky entitled "Genetics and the Origin of Species." The title of this review paper was intended to give readers the (valid) impression that another conceptual and empirical update was needed in evolutionary biology, based this time on new phylogenetic reasoning.

ABSTRACT

A recent criticism that the biological species concept (BSC) unduly neglects phylogeny is examined under a novel modification of coalescent theory that considers multiple, sex-defined genealogical pathways through sexual organismal pedigrees. A competing phylogenetic species concept (PSC) also is evaluated from this vantage. Two analytical approaches are employed to capture the composite phylogenetic information contained within the braided assemblages of hereditary pathways of a pedigree: (i) consensus phylogenetic trees across allelic transmission routes and (ii) composite phenograms from quantitative values of organismal coancestry. Outcomes from both approaches demonstrate that the supposed sharp distinction between BSC and PSC is illusory. Historical descent and reproductive ties are related aspects of phylogeny and jointly illuminate biotic discontinuity.

Sampling properties of genealogical pathways underlying population pedigrees

Wollenberg, K. and J.C. Avise. 1998. Sampling properties of genealogical pathways underlying population pedigrees. **Evolution 52:957–966.**

ANECDOTE OR BACKDROP

Most of JCA's students have had organismal or field-biology backgrounds, but Kurt Wollenberg was far more comfortable at the computer. Accordingly, his dissertation research—part of which is summarized here—dealt with the theory of multiple gene trees in population pedigrees.

ABSTRACT

In sexual species, autosomal alleles are transmitted through multigeneration organismal pedigrees via pathways of descent involving both genders. Here, models assess the sampling properties of these gender-described transmission pathways. An isolation-by-distance model of mating was used to construct a series of computer population pedigrees by systematically varying neighborhood size and the timing of isolation events in sundered populations. For each known pedigree, a matrix of true coancestry coefficients between all individuals in the final generation was calculated and compared (using cophenetic correlations) to mean pairwise times to common ancestry as estimated by sampling varying numbers of gender-defined lineage routes available to individual alleles through that pedigree. When few lineage routes were sampled, agreement between the estimated and the true pedigree was poor and showed a large variance. Agreement improved as more lineage routes were incorporated and asymptotically approached plateau levels that were predictably relatable to the magnitude of population structure. Results underscore a distinction between the composite information in a population pedigree and the subsets of that information registered in allelic lineage pathways.

A comparative summary of genetic distances in the vertebrates from the mitochondrial cytochrome *b* gene

Johns, G.C. and J.C. Avise. 1998. A comparative summary of genetic distances in the vertebrates from the mitochondrial cytochrome b gene. Molecular Biology and Evolution 15:1481−1490.

ANECDOTE OR BACKDROP

The conservative pattern of protein evolution in birds vis-à-vis many other organisms have already been emphasized (see Chapter 12*), but questions remained as to whether such evolutionary conservatism might extend to the DNA level and to other genomes (notably mitochondrial) as well. This paper began to address that issue.*

ABSTRACT

Mitochondrial cytochrome *b* (*cytb*) is among the most extensively sequenced genes to date across the vertebrates. Here we employ nearly 2000 *cytb* gene sequences from GenBank to calculate and compare levels of genetic distance between sister species, congeneric species, and confamilial genera within and across the major vertebrate taxonomic classes. The results of these analyses parallel and reinforce some of the principal trends in genetic distance estimates previously reported in a summary of the multilocus allozyme literature. In particular, surveyed avian taxa on average show significantly less genetic divergence than do same-rank taxa surveyed in other major vertebrate groups, notably reptiles and amphibians. Various biological possibilities and taxonomic "artifacts" are considered that might account for this pattern. Regardless of the explanation, by the yardstick of genetic divergence in this mtDNA gene, as well as genetic distances in allozymes, there is rather poor equivalency of taxonomic rank across some of the vertebrates.

Molecular population structure and the biogeographic history of a regional fauna: a case history with lessons for conservation biology

Avise, J.C. 1992. Molecular population structure and the biogeographic history of a regional fauna: a case history with lessons for conservation biology. Oikos 63:62−76.

ANECDOTE OR BACKDROP

By the early 1990s, JCA and his graduate students and colleagues had accumulated phylogeographic data for many freshwater, terrestrial, and maritime vertebrates and invertebrates sampled across the southeastern United States. This review article pulled all of that genetic information together in a comparative context and emphasized the conservation implications.

ABSTRACT

Mitochondrial (mt) DNA data on the comparative phylogeographic patterns of 19 species of freshwater, coastal, and marine species in the southeastern United States are reviewed. Nearly all assayed species exhibit extensive mtDNA polymorphism, although still orders-of-magnitude less than predicted under neutrality theory if evolutionary effective population sizes of females are similar to current-day census sizes. In both the freshwater and marine realms, deep and geographically concordant forks in intraspecific mtDNA phylogenies commonly distinguish regional populations in the Atlantic versus Gulf Coast areas. These concordant phylogeographic patterns among independently evolving species provide evidence of similar vicariant histories of population separation and can be related tentatively to episodic changes in environmental conditions during the Pleistocene. However, the heterogeneity of observed genetic distances and inferred separation times is difficult to accommodate under a uniform molecular clock. Additional population genetic structure within geographic regions is evidenced by species-specific shifts in frequencies of more closely related mtDNA haplotypes and by high frequencies of private haplotypes in some species. The magnitude of local population structure appears partially related to the life history pattern and dispersal capability of each species. The mtDNA results indicate that conspecific populations can be structured at a wide variety of evolutionary depths. The deeper subdivisions in an intraspecific phylogeny reflect the major sources of evolutionary gene pool diversity within a species, while the shallower molecular separations evidence more recent population subdivisions that can be related to comparative dispersal and gene flow patterns. These molecular findings are relevant to an understanding of biogeographic diversity, and they carry implications for conservation biology.

Speciation durations and Pleistocene effects on vertebrate phylogeography

Avise, J.C., D. Walker, and G.C. Johns. 1998. Speciation durations and Pleistocene effects on vertebrate phylogeography. **Proceedings of the Royal Society of London B 265:1707–1712.**

ANECDOTE OR BACKDROP

Although JCA had long been interested in quantifying and understanding the genetic differences that accrue during speciation events, it was not until the late 1990s that he thought of framing the general problem in a slightly different way, as follows: What is the temporal duration of a typical geographical speciation "event"? This paper offers what we think is a reasonable empirical answer to that query, at least for vertebrate taxa.

ABSTRACT

An approach previously applied to avian congeners is extended in this paper to other vertebrate classes to evaluate Pleistocene phylogeographic effects and to estimate temporal spans of the speciation process (speciation durations) from mitochondrial (mt) DNA data on extant taxa. Provisional molecular clocks are used to date population separations and to bracket estimates of speciation durations between minimum and maximum values inferred from genetic distances between, respectively, extant pairs of intraspecific phylogroups and sister species. Comparisons of genetic distance trends across the vertebrate classes reveal the following: (i) speciation durations normally entail at least 2 million years on average; (ii) for mammals and birds, Pleistocene conditions played an important role in initiating phylogeographic differentiation among now-extant conspecific populations as well as in further sculpting preexisting phylogeographic variety into many of today's sister species; and (iii) for herpetofauna and fishes, inferred Pleistocene biogeographic influences on present-day taxa differ depending on alternative but currently plausible mtDNA rate calibrations.

Species realities and numbers in sexual vertebrates: perspectives from an asexually transmitted genome

Avise, J.C. and D. Walker. 1999. Species realities and numbers in sexual vertebrates: perspectives from an asexually transmitted genome. **Proceedings of the National Academy of Sciences USA** *96:992–995.*

ANECDOTE OR BACKDROP

Few topics in evolutionary biology have been more contentious or received more discussion than species concepts. Are species real? And if so, how do they originate and by what criteria can they be recognized? Ever since early in the twentieth century, most biologists have accepted the biological species concept in which each sexual species is perceived as a reproductive community separated from other such communities by barriers to successful interbreeding. This paper takes a new empirical look at the age-old species issue. Specifically, it asks the following question: Does an asexually inherited molecule (mtDNA) demarcate the same biological entities in nature as do more traditional criteria for species recognition?

ABSTRACT

A literature review is conducted on phylogenetic discontinuities in mtDNA sequences of 252 taxonomic species of vertebrates. About 140 of these species (56%) were subdivided clearly into 2 or more highly distinctive matrilineal phylogroups, the vast majority of which were localized geographically. However, only a small number (two to six) of salient phylogeographic subdivisions (those that stand out against mean within-group divergences) characterized individual species. A previous literature summary showed that vertebrate sister species and other congeners also usually have pronounced phylogenetic distinctions in mtDNA sequence. These observations, taken together, suggest that current taxonomic species often agree reasonably well in number (certainly within an order-of-magnitude) and composition with biotic entities registered in mtDNA genealogies alone. In other words, mtDNA data and traditional taxonomic assignments tend to converge on what therefore may be "real" biotic units in nature. All branches in mtDNA phylogenies are nonanastomose, connected strictly via historical genealogy. Thus, patterns of historical phylogenetic connection may be at least as important as contemporary reproductive relationships *per se* in accounting for micro-evolutionary unities and discontinuities in sexually reproducing vertebrates. Findings are discussed in the context of the biological and phylogenetic species concepts.

ADDENDUM

Most biologists now acknowledge that species in nature tend to be real (rather than artifactual) biological entities.

Proposal for a standardized temporal scheme of biological classification for extant species

Avise, J.C. and G.C. Johns. 1999. Proposal for a standardized temporal scheme of biological classification for extant species. **Proceedings of the National Academy of Sciences USA** *96:7358–7363.*

ANECDOTE OR BACKDROP

In 1966, a book by a formerly rather obscure German entomologist was translated into English, and it soon became a Bible for the emerging cladistic school of systematics. Willi Hennig's "Phylogenetic Systematics" revolutionized the field by advancing the notion that classifications should reflect the branching patterns in phylogenetic trees. Hennig's treatise contained many conceptual insights, not the least being his seminal distinction between synapomorphies (shared-derived traits) and symplesiomorphies (shared-ancestral traits). Many systematists were captivated by Hennig's ideas, and they soon launched a cladistic revolution that they sometimes have proselytized with a religious-like fervor. Strangely, however, cladists have totally neglected another of Hennig's suggestions: that phylogenetic classifications should be standardized such that the categorical rank of each taxon would denote its geological age. This paper was our effort to resurrect this interesting proposal, which we suspect could actually be brought to fruition (if systematists decided to do so) in the modern era of geological dating by molecular clocks.

ABSTRACT

With respect to conveying useful comparative information, current biological classifications are seriously flawed because they fail to: (i) standardize criteria for taxonomic ranking and (ii) equilibrate assignments of taxonomic rank across disparate kinds of organisms. In principle, these problems could be rectified by adopting a universal taxonomic yardstick based on absolute dates of the nodes in evolutionary trees. By using procedures of temporal banding described herein, a simple philosophy of biological classification is proposed that would retain a manageable number of categorical ranks yet apply them in standardized fashion to time-dated phylogenies. The phylogenetic knowledge required for a time-standardized nomenclature arguably may emerge in the foreseeable future from vast increases in multi-locus DNA sequence information (coupled with continued attention to phylogeny estimation from traditional systematic data). By someday encapsulating time-dated phylogenies in a familiar yet modified hierarchical ranking scheme, a temporal banding approach would improve the comparative information content of biological classifications.

ADDENDUM

This taxonomic proposal by JCA and Hennig seems to have been almost universally rejected (or at least ignored) by the systematics community.

Cytonuclear genetic signatures of hybridization phenomena: rationale, utility, and empirical examples from fishes and other aquatic animals

Avise, J.C. 2001. Cytonuclear genetic signatures of hybridization phenomena: rationale, utility, and empirical examples from fishes and other aquatic animals. Reviews in Fish Biology and Fisheries 10:253–263.

ANECDOTE OR BACKDROP

This was an invited review article in which JCA was asked to summarize his laboratory's work on two fronts: (i) the development of cytonuclear disequilibrium statistics (see Chapter 8*) and (ii) their application to the study of hybrid zones in fishes and other aquatic creatures. This overview paper was the result of that request.*

ABSTRACT

By definition, organisms of hybrid ancestry carry amalgamations of divergent genomes. Thus, exaggerated effects of genomic interactions might be anticipated in hybrid populations, thereby magnifying the impact of natural selection and making this and other evolutionary forces easier to document. Mating biases and other gender-based asymmetries also frequently characterize hybrid populations. Thus, maternally inherited cytoplasmic polymorphisms assayed jointly with those at biparentally inherited nuclear loci provide powerful genetic markers to dissect ethological, ecological, and evolutionary processes in hybrid settings. Population-level topics that can be addressed using cytonuclear markers include the frequency of hybridization and introgression in nature, behavioral and ecological factors (such as mating preferences and hybrid fitnesses) influencing the genetic architectures of hybrid zones, the degree of consistency in genetic outcomes across multiple hybrid contact regions, and environmental impacts (including the introduction of alien species) on hybridization processes. Several empirical studies on fish populations in hybrid settings illustrate the application of cytonuclear appraisals in such contexts.

Genetic perspectives on the natural history of fish mating systems

DeWoody, J.A. and J.C. Avise. 2001. Genetic perspectives on the natural history of fish mating systems. **Journal of Heredity 92:167–172.**

ANECDOTE OR BACKDROP

By the late 1980s, postdoc Andrew DeWoody and others in JCA's laboratory had conducted detailed genetic dissections of biological parentage (paternity and maternity) and mating systems in a wide variety of nest-tending fish species (see Chapters 2, 4, and 6). *This paper reviewed such studies.*

ABSTRACT

Molecular analyses of bird and mammal populations have shown that social mating systems must be distinguished from genetic mating systems. This distinction is important in fishes also, where the potential for extra-pair spawning and intraspecific brood parasitism is especially great. We review studies on freshwater and marine fishes that have used molecular markers to document biological parentage and genetic mating systems in nature, particularly in species with extended parental care of offspring. On average, nest-guarding adults parented about 70–95% of their custodial offspring, and approximately one-third of the nests were cuckolded to some extent. Furthermore, nearly 10% of the assayed nests contained offspring tended by foster fathers either because of nest takeovers or egg thievery. On average, fish that provide parental care on nests spawned with more mates than did fish with internal fertilization and pregnancy. Overall, genetic markers have both confirmed and quantified the incidence of several reproductive and other social behaviors of fishes, and have thereby enhanced our knowledge of piscine natural history.

Genetic mating systems and reproductive natural histories of fishes: lessons for ecology and evolution

Avise, J.C., A.G. Jones, D. Walker, J.A. DeWoody, B. Dakin, A. Fiumera, D. Fletcher, M. Mackiewicz, D. Pearse, B. Porter, and S.D. Wilkins. 2002. Genetic mating systems and reproductive natural histories of fishes: lessons for ecology and evolution. **Annual Review of Genetics 36:19–45.**

ANECDOTE OR BACKDROP

This was a more extended overview of our genetic findings on the reproductive natural histories of diverse piscine taxa.

ABSTRACT

Fish have diverse breeding behaviors that make this taxonomic group valuable for testing theories on the ecology and evolution of genetic mating systems and alternative reproductive tactics. Here we review DNA level appraisals of paternity and maternity in wild fish populations, and integrate the molecular data with natural history information. Behavioral phenomena unveiled and quantified in various species by the genetic markers include: multiple mating by either or both genders, frequent cuckoldry by males and occasional cuckoldry by females in nest-tending species, additional routes to foster parentage via nest piracy and egg thievery, brood parasitism by helper males in cooperative nesters, clutch mixing in oral brooders, filial cannibalism, kin associations (or lack thereof) in schooling fry of broadcast spawners, sperm storage by dams in female pregnant species, and enhanced sexual selection on females in some male pregnant species. These and other results from the genetic parentage analyses are discussed in the context of relevant behavioral theory.

Microsatellite null alleles in parentage analysis

Dakin, E.E. and J.C. Avise. 2004. Microsatellite null alleles in parentage analysis. **Heredity 93:504–509.**

ANECDOTE OR BACKDROP

In many of our genetic analyses of paternity and maternity in natural populations (see several earlier chapters), an annoying complication was the presence of null alleles that fail to amplify under the PCR conditions employed. Here JCA's student Beth Dakin reviewed this phenomenon in the broader literature.

ABSTRACT

Highly polymorphic microsatellite markers are widely employed in population genetic analyses (e.g., of biological parentage and mating systems), but one potential drawback is the presence of null alleles that fail to amplify to detected levels in the PCR assays. Here we examine 233 published articles in which authors reported the suspected presence of one or more microsatellite null alleles, and we review how these purported nulls were detected and handled in the data analyses. We also employ computer simulations and analytical treatments to determine how microsatellite null alleles might impact molecular parentage analyses. Results indicate that whereas null alleles in frequencies typically reported in the literature introduce rather inconsequential biases on average exclusion probabilities, they can introduce substantial errors into empirical assessments of specific mating events by leading to high frequencies of false parentage exclusions.

Phylogenetic perspectives on the evolution of parental care in ray-finned fishes

Mank, J.E, D.E.L. Promislow, and J.C. Avise. 2005. Phylogenetic perspectives on the evolution of parental care in ray-finned fishes. Evolution 59:1570—1578.

ANECDOTE OR BACKDROP

In the early 2000s, a brilliantly talented and energetic young woman began working toward her Ph.D. in JCA's laboratory. Judith Mank first canvassed the extensive scientific literatures on fish systematics and piscine reproductive behaviors, and then wedded these two topics in an evolutionary context. Her basic approach was to estimate a phylogenetic supertree for bony fishes and then map behavioral or other sexual features onto that tree in order to deduce the evolutionary histories of various reproductive traits. This approach involving PCM (phylogenetic character mapping) was a nice complement to the lab's longstanding empirical focus on using molecular markers to unveil reproductive modes and behaviors in particular fish taxa. This paper and the three that follow are representative of a long series of PCM analyses that Judith produced as a part of her lengthy dissertation.

ABSTRACT

Among major vertebrate groups, ray-finned fishes (Actinopterygii) collectively display a nearly unrivaled diversity of parental care activities. This fact, coupled with a growing body of phylogenetic data for Actinopterygii, makes these fishes a logical model system for analyzing the evolutionary histories of alternative parental care modes and associated reproductive behaviors. From an extensive literature review, we constructed a supertree for ray-finned fishes and used its phylogenetic topology to investigate the evolution of several key reproductive states including type of parental care (maternal, paternal, or biparental), internal versus external fertilization, internal versus external gestation, nest construction behavior, and presence versus absence of sexual dichromatism (as an indicator of sexual selection). Using a comparative phylogenetic approach, we critically evaluate several hypotheses regarding evolutionary pathways toward parental care. Results from maximum parsimony reconstructions indicate that all forms of parental care, including paternal, biparental, and maternal (both external and internal to the female reproductive tract) have arisen repeatedly and independently during ray-finned fish evolution. The most common evolutionary transitions were from external fertilization directly to paternal care, and from external fertilization to maternal care via the intermediate step of internal fertilization. We also used maximum likelihood phylogenetic methods to test for statistical correlations and contingencies in the evolution of pairs of reproductive traits. Sexual dichromatism and nest construction proved to be positively correlated with the evolution of male parental care in species with external fertilization. Sexual dichromatism was also positively correlated with female-internal fertilization and gestation. No clear indication emerged that female-only care or biparental care were evolutionary outgrowths of male-only care, or that biparental care has been a common evolutionary stepping-stone between paternal and maternal care. Results are discussed in the context of prior thought about the evolution of alternative parental care modes in vertebrates.

Sex chromosomes and male ornaments: a comparative evaluation in ray-finned fishes

Mank, J.E., D.W. Hall, M. Kirkpatrick, and J.C. Avise. 2006. Sex chromosomes and male ornaments: a comparative evaluation in ray-finned fishes. **Proceedings of the Royal Society of London B 273:233–236.**

ANECDOTE OR BACKDROP

The idea behind a publication is sometimes (indeed, usually) sparked by some serendipitous event. This paper is a case in point. The study was initiated when a visiting speaker—Mark Kirkpatrick— delivered a seminar on sexual selection at the University of Georgia. At that time, Judith Mank was completing her dissertation in JCA's lab, and we realized that some of her data might be suitable for testing an intriguing hypothesis that Mark had raised. Several months later, this paper emerged from our chance collaboration.

ABSTRACT

Theory predicts that the mechanism of genetic sex determination can substantially influence the evolution of sexually selected traits. For example, female heterogamety ($\male$ZZ/$\female$ZW) can favor the evolution of extreme male traits under Fisher's runaway model of sexual selection. We empirically test whether the genetic system of sex determination has played a role in the evolution of exaggerated male ornaments in actinopterygiian fishes, a clade in which both female heterogametic and male heterogametic systems of sex determination have evolved multiple times. Using comparative methods both uncorrected and corrected for phylogenetic nonindependence, we detected no significant correlation between sex chromosome systems and sexually selected traits in males. Results suggest that sex determination mechanism is at best a relatively minor factor affecting the outcomes of sexual selection in ray-finned fishes.

Comparative phylogenetic analysis of male alternative reproductive tactics in ray-finned fishes

Mank, J.E. and J.C. Avise. 2006. Comparative phylogenetic analysis of male alternative reproductive tactics in ray-finned fishes. **Evolution 60:1311–1316.**

ABSTRACT

Using comparative phylogenetic analysis, we analyzed the evolution of male alternative reproductive tactics (MARTs) in ray-finned fishes (Actinopterygii). Numerous independent origins for each type of MART (involving sneaker males, female mimics, pirates, and satellite males) indicate that these behaviors have been highly labile across actinopterygiian evolution, consistent with a previous notion that convergent selection in fishes can readily mold the underlying suites of reproductive hormones into similar behaviors. The evolutionary appearance of MARTs was significantly correlated with the presence of sexually selected traits in bourgeois males ($p = 0.001$) but not with the presence of male parental care. This suggests that MARTs often arise from selection on some males to circumvent bourgeois male investment in mate monopolization, rather than to avoid male brood care *per se*. We found parsimony evidence for an evolutionary progression of MARTs wherein sneaking is usually the evolutionary precursor to the presumably more complex MARTs of female mimicry and cooperative satellite behavior. Nest piracy appears not to be part of this evolutionary progression, possibly because its late onset in the life cycle of most ray-finned fishes reduces the effects of selection on this reproductive tactic.

Evolution of alternative sex-determining mechanisms in Teleost fishes

Mank, J.E., D.E.L. Promislow, and J.C. Avise. 2006. Evolution of alternative sex-determining mechanisms in Teleost fishes. **Biological Journal of the Linnean Society** *87:83–93.*

ABSTRACT

We compile information from the literature on the taxonomic distributions in extant teleost fishes of alternative sex determination systems: male heterogametic (XY) gonochorism, female heterogametic (ZW) gonochorism, hermaphroditism, unisexuality, and environmental dependency. Then, using recently published molecular phylogenies based on whole genomic or partial mitochondrial DNA sequences, we infer the histories and evolutionary transitions between these reproductive modes by employing maximum parsimony and maximum likelihood methods of phylogenetic character mapping. Across a broad teleost phylogeny involving 25 taxonomic orders, a highly patchy distribution of different sex determination mechanisms was uncovered, implying numerous transitions between alternative modes, but this heterogeneity also precluded definitive statements about ancestral states for most clades. Closer inspection of family-level and genus-level phylogenies within each of four orders further bolstered the conclusion that shifts in sex-determining modes are evolutionarily frequent and involve a variety of distinct ancestral-descendant pathways. For possible reasons discussed herein, the evolutionary lability of sex-determining modes in fishes contrasts strikingly with the evolutionary conservatism of sex determination within both mammals and birds.

Time to standardize taxonomies

Avise, J.C. and D. Mitchell. 2007. Time to standardize taxonomies. **Systematic Biology** *56:130–133.*

ANECDOTE OR BACKDROP

Another serendipitous event led to this coauthored paper. It all began 1 day when JCA received an email from someone he did not know. Dale Mitchell's letter of introduction explained that he was an amateur science buff who had been intrigued by a publication of mine arguing for greater standardization in systematics (see "Proposal for a standardized temporal scheme..." earlier in this chapter). Dale went on in his letter to state that although he liked the idea, the "temporal banding" approach that JCA had advocated would create undue havoc in biology by necessitating name changes for many taxa. Dale asked "Why not instead just attach a short time clip, denoting the geological age of clade origin, to the conventional name for each taxon? The idea was simple yet elegant! So, JCA quickly sent to Dale a draft manuscript that is summarized in this abstract.

ABSTRACT

Systematists for years have engaged in debates about whether to retain, overhaul, or abandon conventional Linnaean frameworks for biological classification. Our time-clip amendment to the original temporal banding proposal provides a practical way to retain the familiar Linnaean system *and* simultaneously promotes the incorporation of new phylogenetic discoveries from molecular biology, paleontology, or other relevant evolutionary disciplines. By enabling biologists to signify and sort traditional Linnaean taxa according to their approximate dates of origin, time clips would add materially to the information content of biological classifications, help to equilibrate nonstandardized taxonomic ranks across disparate forms of life, promote novel avenues of thought about the comparative evolutionary rates of organismal phenotypes and genotypes, and in general facilitate nearly all types of scientific research in phylogenetics.

On the temporal inconsistencies of Linnean taxonomic ranks

Avise, J.C. and J.-X. Liu. 2011. On the temporal inconsistencies of Linnean taxonomic ranks. **Biological Journal of the Linnean Society 102:707–714.**

ANECDOTE OR BACKDROP

This study was a follow-up to the previous one. It empirically documents that there is indeed a problem related to the lack of consistence and comparability in conventional Linnean classifications.

ABSTRACT

The inconsistency problem in systematics refers in part to the fact that disparate taxa of identical Linnean rank are not necessarily similar or even readily comparable in any other specifiable biological feature. This shortcoming led to a "temporal banding" proposal in which extant clades associated with particular taxonomic ranks would be standardized according to a universal metric: the absolute time of evolutionary origin. However, one underexplored possibility is that same-level taxa in disparate organismal groups might already be similar (fortuitously so) in evolutionary age. In the present study, we explicitly address this possibility by reviewing published molecular inferences about the known or suspected origination dates of taxonomic genera, families, and orders in diverse organismal groups. Our findings empirically confirm that currently recognized taxa are far from temporally standardized, thereby adding support for the contention that this kind of taxonomic inconsistency should ultimately be rectified in our biological classifications.

Multiple mating and its relationship to alternative modes of gestation in male-pregnant versus female-pregnant fish species

Avise, J.C. and J.-X. Liu. 2010. Multiple mating and its relationship to alternative modes of gestation in male-pregnant versus female-pregnant fish species. **Proceedings of the National Academy of Sciences USA** *107:18915–18920.*

ANECDOTE OR BACKDROP

This paper and the two that follow provide empirical overviews of genetic data and theory for three different contexts of "pregnancy," respectively: (i) male pregnancy (see Chapter 5*) versus female pregnancy (see* Chapter 3*) in fishes; (ii) pregnancy in fishes versus the phenomenon in mammals; and (iii) various forms of pregnancy in vertebrates versus their analogues (parental brooding) in invertebrates. All three studies were conducted in collaboration with postdoc Jason Liu from the Peoples Republic of China.*

ABSTRACT

We construct a verbal and graphical theory (the "fecundity-limitation hypothesis") about how constraints on the brooding space for embryos probably truncate individual fecundity in male pregnant and female pregnant species in ways that should differentially influence selection pressures for multiple mating by males or by females. We then review the empirical literature on genetically deduced rates of multiple mating by the embryo-brooding parent in various fish species with three alternative categories of pregnancy: internal gestation by males, internal gestation by females, and external gestation (in nests) by males. Multiple mating by the brooding gender was common in all three forms of pregnancy. However, rates of multiple mating as well as mate numbers for the pregnant parent averaged higher in species with external as compared to internal male pregnancy and also for dams in female pregnant species versus sires in male pregnant species. These outcomes are all consistent with the theory that different types of pregnancy have predictable consequences for a parent's brood space, its effective fecundity, its opportunities and rewards for producing half-sib clutches, and thereby its exposure to selection pressures for seeking multiple mates. Overall, we try to fit these fecundity-limitation phenomena into a broader conceptual framework for mating system evolution that also includes anisogamy, sexual selection gradients, parental investment, and other selective factors that can influence the relative proclivities of males versus females to seek multiple sexual partners.

Multiple mating and its relationship to brood size in pregnant fishes versus pregnant mammals and other viviparous vertebrates

Avise, J.C. and J.-X. Liu. 2011. Multiple mating and its relationship to brood size in pregnant fishes versus pregnant mammals and other viviparous vertebrates. **Proceedings of the National Academy of Sciences USA 108:7091−7095.**

ABSTRACT

We summarize the genetic literature on rates of multiple paternity and sire numbers per clutch in viviparous fishes versus mammals, two vertebrate groups in which the pregnancy phenomenon is common but entails very different numbers of embryos (for the species surveyed, piscine broods averaged >10-fold larger than mammalian litters). As deduced from the genetic parentage analyses, multiple mating by the pregnant gender proved to be common in most of the assayed species but averaged significantly higher in fishes than in mammals. However, *within* either of these two vertebrate groups we found no significant positive correlations between brood size and genetically deduced incidences of multiple mating by females. Overall, these findings offer little support for the hypothesis that clutch size in pregnant species closely predicts the outcome of selection for multiple mating by the brooders. Instead, whatever factors promote multiple mating by members of the gestating gender seem to do so in surprisingly similar ways in live-bearing vertebrates otherwise as different as fish and mammals. Similar conclusions emerged when we extended the survey to include viviparous species of amphibians and reptiles. One notion consistent with all of these empirical observations is that although several fitness benefits probably accrue from multiple mating, logistical constraints on mate-encounter rates routinely truncate multiple mating far below levels that otherwise could be accommodated, especially in species with larger broods. We develop this concept into a "logistical constraint hypothesis" (LCH) that may help to explain these mating outcomes in viviparous vertebrates. Under the LCH, propensities for multiple mating in each species register a balance between near-universal fitness benefits from multiple mating and species-idiosyncratic logistical limits on polygamy.

Multiple mating and clutch size in invertebrate brooders versus pregnant vertebrates

Avise, J.C., A. Tatarenkov, and J.-X. Liu. 2011. Multiple mating and clutch size in invertebrate brooders versus pregnant vertebrates. **Proceedings of the National Academy of Sciences USA 108:11512−11517.**

ABSTRACT

We summarize the genetic literature on polygamy rates and sire numbers per clutch in invertebrate animals that brood their offspring and then compare findings to analogous data previously compiled for vertebrate species displaying viviparity or other pregnancy-like syndromes. As deduced from molecular parentage analyses of several thousand broods from more than 100 "pregnant" species, invertebrate brooders had significantly higher mean incidences of multiple mating than pregnant vertebrates, a finding generally consistent with the postulate that clutch size constrains successful mate numbers in species with extended parental care. However, we uncovered no significant correlation in invertebrates between brood size and genetically deduced rates of multiple mating by the incubating sex. Instead, in embryo-gestating animals otherwise as different as mammals and mollusks, polygamy rates and histograms of successful mates per brooder proved to be strikingly similar. Whereas most previous studies have sought to understand why gestating parents have so many mates and such high incidences of successful multiple mating, an alternative perspective based on logistical constraints turns the issue on its head by asking why mate numbers and polygamy rates are much lower than they theoretically could be given the parentage-resolving power of molecular markers and the huge sizes of many invertebrate broods.

ADDENDUM

The themes developed in these last three papers (see also Chapters 2—6) *were a part of the motivation for JCA's recent (2013) book entitled "Evolutionary Perspectives on Pregnancy."*

Epilogue

The Golden Era of molecular ecology and evolution during the final third of the twentieth century witnessed scientific transformations that will never be repeated. For the first time, biologists gained access to molecular data from polymorphic DNA sequences and the proteins that they often encode. For the first time, an entire generation of scientists that had been classically trained in ecology, organismal biology, and systematics was literally forced to adjust to a new way of looking at the natural world. For the first time, all of nature was opened for close genetic scrutiny. This Golden Age was a time when almost any species— no matter how charismatic—could be genetically examined with some assurance that the creature had never before been examined thusly. It was a time when organismal biology and molecular genetics first became fused into a grand new evolutionary synthesis. Now that we have successfully traversed the original Golden Age of molecular ecology and evolution, I can foresee the dawn of a new genetic revolution as we enter the modern genomics era. In the near and not-too-distant future, I predict that the broader emphasis will shift from the use of molecules as mere markers for ecology and natural history, to a more explicit and comprehensive search for the mechanistic genetic underpinnings of numerous organismal phenotypes. Thus, in another 50 years we may well be able to reflect back upon a second Golden Era for the fields of molecular ecology and evolution.

In the meantime, I hope that this book has captured the essence of one scientific revolution—as viewed through the prism of one of its most fascinated participants—in a manner that is both informative to professional biologists and aesthetically pleasing to almost anyone.

Index

Printed in the United States
By Bookmasters